Atomic and Molecular Physics
(Second Edition)

A primer

Online at: https://doi.org/10.1088/978-0-7503-5734-0

Atomic and Molecular Physics (Second Edition)

A primer

Luciano Colombo

Department of Physics, University of Cagliari, Italy

IOP Publishing, Bristol, UK

ISBN   978-0-7503-5734-0 (ebook)
ISBN   978-0-7503-5732-6 (print)
ISBN   978-0-7503-5735-7 (myPrint)
ISBN   978-0-7503-5733-3 (mobi)

DOI   10.1088/978-0-7503-5734-0

Version: 20231001

IOP ebooks

British Library Cataloguing-in-Publication Data: A catalogue record for this book is available from the British Library.

Published by IOP Publishing, wholly owned by The Institute of Physics, London

IOP Publishing, No.2 The Distillery, Glassfields, Avon Street, Bristol, BS2 0GR, UK

US Office: IOP Publishing, Inc., 190 North Independence Mall West, Suite 601, Philadelphia, PA 19106, USA

To my family.

Contents

Part II Atomic physics

3 One-electron atoms   3-1

Part III Molecular physics

Part IV Concluding remarks

Part V Appendices

Introduction to the second revised and extended edition

This second edition comes four years after the first. In these years the textbook has been used by the students attending my bachelor course on condensed matter physics given at the University of Cagliari, as well as being widely used both in Italy and abroad. Accordingly, over these years I have received many reports from careful readers pointing out misprints, inaccuracies in the figures and, in some cases (fortunately just a few!), errors in the formulae or in their derivation. As a first task, therefore, I carefully reviewed the entire text of the manual, correcting where necessary.

I also received stimulating comments on how to better explain a given concept, or how to develop in a more accessible way some technical passages that were a little more difficult, either for conceptual reasons or for mathematical complication. This, in particular, concerned:

1. the taxonomy of relativistic effects (chapter 3), with a special note on spin (which is now proved not to be a classical degree of freedom) and the Lamb-shift effect;
2. the microscopic theory of Einstein coefficients (chapter 4);
3. the electronic structure of polyatomic molecules (chapter 8).

In both cases, I entirely rewrote the relevant sessions, adding new material and developing the formalism to a more accurate and complete level.

I also decided to complete the textbook by adding some new topics and figures. My intention was to improve the coverage of atomic and molecular physics by making it more complete, but without altering the tutorial character of the textbook and, therefore, still preserving its deliberately non-encyclopedic nature (in other words, the 'primer' characteristics have been preserved, according to the concepts described in the Introduction to the first edition). In more detail:

1. I discussed the finite width of the spectral lines;
2. I have greatly expanded the section devoted to the LASER, providing much more information about the operation of such a device;
3. I have explicitly treated the case of alkaline atoms which, together with the helium atom case-study (already present in the previous edition), provides a useful introduction to the physics of multi-electron atoms;
4. I have included a new section on excited states in multi-electron atoms (in the first edition, the discussion was limited just to their ground state);
5. I have thoroughly discussed the rudiments of the Hückel method, a semi-empirical theory that can be straightforwardly developed with full rigour to predict the electronic structure of polyatomic molecules.

Finally, I have added two new appendices: as already commented in the first edition, their reading can be skipped without compromising the understanding of what is presented in the main chapters; however, they represent the ideal completion and deepening of what is there discussed. In particular, the new edition is enriched with appendices devoted to:

1. the evaluation of effects due to the finite size of the atomic nucleus;
2. a simple derivation of classical Boltzmann distribution, referred to on several occasions in the text.

Finally, the appendix describing the minimum coupling scheme has been extensively revised, making it more rigorous.

My revision and extension work has greatly benefited from the help of many. First of all, I would like to thank all my students who, by pointing out errors or inaccuracies, have motivated and stimulated me to readdress the content of my textbook. Next, I am especially grateful to Dr Antonio Cappai (University of Cagliari) not only for his careful and critical reading of the new edition, but also for his preliminary drafting of some extensions that I eventually inserted in new sections or appendices. Finally, I acknowledge many useful discussions with Mr Robert Panai (University of Cagliari).

Perfection is not of this world and, therefore, even this second revised and extended edition of 'Atomic and Molecular Physics: a primer' may contain inaccuracies, for which I am solely responsible.

LUCIANO COLOMBO
Department of Physics
University of Cagliari
Cagliari, May 2023

Foreword to the first edition

There are many excellent books teaching physics at different levels of difficulty. However, if we look not only for good physics but *also* for attractive and pleasing ways to present the challenging physical questions, as well as the ways to face and solve them, we soon come to the conclusion that, except for some celebrated cases, there are not many. Yet, to feel that what is being learned is something which is not really that difficult but something natural or even relatively simple, is what makes a reader feel comfortable with new notions and, in the end, drives him/her to rethink these ideas and thus, to discover that in fact things are not that simple and what is being read is a clever answer to a complex problem. But this is the way for many of us to learn, as in one of these Chekhov plays so full of real-life insight although apparently nothing extraordinary or unusual happens. We must feel at home with new concepts to experience the need to go to the heart of the problem. The gift to be able to explain complex concepts in simple words is the result of a long practice in polishing from unnecessary complications and reducing the problem to its very fundamental nature. Professor Colombo is an accomplished communicator as any reader of this book will agree. The present volume, which is the first of a series of three aiming to present a broad coverage of atomic, molecular, solid-state and statistical physics, typically at an undergraduate level, will be enjoyed by everyone trying to enter into the realm of atoms and molecules and how we understand them.

The first chapter ('The overall picture') already sets the tone. Do atoms exist? What are atoms made of? These are the *simple* questions opening the book. In about one dozen pages we start being told about such fundamental things as the law of multiple definite proportions and Avogadro's number, the different experimental realizations providing support for the atomistic picture, the constituents of atoms and the experimental observations pointing to failures of classical physics when dealing with small enough objects, and eventually leading to the wave–matter duality and the development of the matter waves equation and its probabilistic interpretation. This is really a long trip dealing with difficult notions which could discourage many readers, but which is absorbing and thought-provoking in the author's hands.

The volume consists of three main parts: 'Preliminary concepts', 'Atomic physics' and 'Molecular physics'. The first part contains two chapters which provide the appropriate historical perspective leading to the contemporary view of atomic and molecular physics, as well as the necessary mathematical background. As discussed above, the first chapter is a guided tour through the physical developments which led from the atomistic hypothesis to the shaping of the new language needed to properly describe atoms and molecules based on the wave–matter duality. The second chapter ('Essential quantum mechanics') introduces the basic notions and mathematical tools, i.e., the basic ingredients of the language which will be used in the rest of the book. All over this chapter, which is thus an introduction to essential quantum mechanics, effort is made in underlining how the new language is related to notions in classical physics but also what are the subtle new ideas emerging, as for instance the far-reaching consequences of the concept of indistinguishability of identical particles in quantum mechanics leading to the Pauli principle. This chapter concludes with the introduction of perturbation theory, a wonderful tool which will be extensively used in the two main parts of the book.

The second part ('Atomic physics'), containing three chapters, deals with the physical description of atoms and their interaction with radiation. It begins with application of the machinery developed in chapter 2 to the simplest atomic system, the hydrogen atom. The study leads to the important concept of atomic orbital, a basic ingredient of our knowledge of atomic and molecular physics and chemistry. Atomic orbitals describe the movement of the electron in the field of charged nuclei and this leads naturally to the study of the magnetic interactions occurring in one-electron atomic systems. The effect of uniform and non-uniform magnetic fields is considered and leads to another extremely important concept, the electron spin. This concept is clearly introduced, showing how it can provide an explanation for the famous Stern–Gerlach experiment. A simple, intuitive explanation of its origin is presented although it is already pointed out that only in a relativistic version of quantum mechanics can the spin concept be naturally accounted for. This chapter is completed with the introduction of the spin–orbit coupling effects and the fine and hyperfine structure of one-electron atoms spectra. This is followed by a chapter dealing with the interaction of one-electron atoms with electromagnetic radiation. The elementary theory of absorption and emission of radiation is developed for both spontaneous and stimulated processes and selection rules for allowed transitions are worked out. Finally, the LASER concept is qualitatively presented. The third and last chapter of this part deals with the extension of most of these ideas to many-electron atoms by taking into account electron–electron (Coulomb and exchange) interactions at different levels of sophistication. For instance using the so-called central-field approximation readers are naturally led to the main concepts behind the periodic table of elements. However, more sophisticated approaches to the electron–electron interactions are needed in order to be able to fully understand atomic spectra and the Hartree, Hartree–Fock and configuration interaction methods are discussed. This part concludes with the development of selection rules for optical transitions and the effect of external magnetic fields on multi-electron atoms. It is certainly a long journey full of unexpected surprises but also lucid rationalizations made appealing by the clarity of the exposition.

The third part deals with molecular physics. Now, things become more complex since there are many electrons *and* nuclei. This means that one needs to introduce some approximation in order to keep the problems manageable. This is the Born–Oppenheimer approximation, which is the beginning of the excursion through the molecular world. The rest of this chapter is devoted to the analysis of the two main sources of stability in molecules: ionic and covalent bonding. Whereas ionic bonding can be understood from a purely classical, Coulombic perspective, covalent bonding can only be understood from a purely quantum mechanical approach. The simplest molecule, H_2, is used to discuss this extraordinarily important concept which lies at the basis of modern chemistry. Having set the basis for understanding the electronic motion in diatomic molecules it is time to progress and consider the inner molecular motions, i.e., rotations and vibrations, which in the case of diatomic molecules can be discussed with great accuracy, including the roto-vibrational coupling. In that way the rotational and vibrational spectra used to characterize diatomic molecules is described and later completed by the consideration of light scattering experiments like Raman and Rayleigh scattering. Finally, an extension to polyatomic molecules is presented leading to the introduction of the very useful concept of normal vibrational mode. In the final chapter of this part the electronic structure of polyatomic molecules is discussed. Analysis of the simple H_2^+ molecular ion leads to the introduction of several important concepts like molecular orbital, overlap and resonance integrals, bonding and antibonding levels and electronic configuration, which are at the basis of the modern theory of the chemical bonding. The linear combination of atomic orbitals approach, a formalism which has strong roots in solid-state theory and is very well suited to fully exploit the symmetry properties of the molecules is used throughout the chapter. The nature of the molecular orbitals for different homonuclear and heteronuclear diatomic molecules as well as polyatomic molecules is discussed so that the reader is acquainted with many aspects of the conceptual framework commonly used to discuss the electronic structure of molecular systems (hybridization, electronic delocalization, etc).

This book provides a guide to understanding the physics of atoms and molecules in a vivid and pedagogical way. Throughout the book the mathematical formalism is kept to a minimum level compatible with rigor without leading to unnecessarily long derivations which would distract the reader from the conceptual framework. When such mathematical digressions are useful they are given in a series of appendices.

One could naively think that if we are able to understand something we should also be able to explain it to others. This is unfortunately not so simple. Being able to expose in a clear, consistent and engaging way is something requiring a long dedication. Professor Colombo has wholly succeeded in this purpose and provided us with a nice and attractive approach to atomic and molecular physics. The level of the book is typically that for an undergraduate series of lectures and thus will be well suited for an audience of students in physics, chemistry, materials sciences and engineering having a good background in classical physics and calculus. The good balance between fundamental understanding and formal treatment will also make it appropriate for graduate students or scientists in other fields needing to acquire a good grasp of atomic or molecular physics.

Those who will have their first exposure to atomic and molecular physics through this book are very lucky. There is no doubt that readers of this book will be eagerly waiting for the next two volumes.

ENRIC CANADELL
Institute of Materials Sciences of Barcelona
Consejo Superior de Investigaciones Científicas
Bellaterra, June 2019

Acknowledgements

I am really indebted to Caroline Mitchell (Commissioning Editor of IOP Publishing) and Daniel Heatley (ebooks Editorial Assistant of IOP Publishing) for their great enthusiasm in accepting my initial proposal for the first edition of this Primer and for shaping it into a suitable editorial product.

Now it is a pleasure to warmly acknowledge Emily Tapp (Commissioning Editor of IOP Publishing) for supporting my writing efforts for the second edition of this Primer with great professionalism, always promptly replying to my queries and clarifying my doubts. Her assistance has been truly invaluable.

Presentation of the 'Primer series'

This is the first volume of a series of three books that, as a whole, account for an introduction to the huge field usually referred to as 'condensed matter physics': they are, respectively, addressed to atomic and molecular physics, to solid state physics, and to statistical methods for the description of classical or quantum ensembles of particles. They are based on my 20-year experience of teaching undergraduate courses on these topics for bachelor-level programs in physical and engineering sciences at the University of Cagliari (Italy).

The volumes are called 'Primers' to underline that the pedagogical aspects have been privileged over those of completeness. In particular, I selected the contents of each volume so to keep limited its number of pages and so that the topics actually covered correspond to the *syllabus* of a typical one-semester course.

More important, however, was the choice of the style of presentation: I wanted to avoid an excessively formal treatment, preferring instead the exploration of the underlying physical features and always placing phenomenology at the centre of the discussion. More specifically, the main characteristics of this book series are:

- emphasis is always given to the physical content, rather than to formal proofs, i.e., mathematics is kept to the minimum level possible, without affecting rigour or clear thinking;
- an in-depth analysis is presented about the merits and faults of any approximation used, incorporating as well a thorough discussion of the conceptual framework supporting any adopted physical model;
- prominence is always on the underlying physical basis or principle, rather than to applications;
- when discussing the proposed experiments, the focus is given to their conceptual background, rather than to the details of the instrumental setup.

Despite the tutorial approach, I nevertheless wanted to follow the Italian academic tradition, which provides even the elementary introduction to condensed matter physics at a quantum level. I hope that my efforts have optimally combined ease-of-access and rigour, especially conceptual.

The intentionally non-encyclopaedic content and the tutorial character of these 'Primers' should facilitate their use even for students not specifically enrolled in a university *curriculum* in physics. I hope, in particular, that my textbooks could be accessible to students in chemistry, materials science and also of many engineering branches. In view of this, I have included a brief outline of non-relativistic quantum mechanics in the first 'Primer', a subject that does not appear in the typical engineering *curricula*. For the rest, classical mechanics, elementary thermodynamics and Maxwell theory of electromagnetism are used, to which all students of natural and engineering sciences are normally exposed.

Each 'Primer' is organized in parts, divided into chapters. This structure is tailored to facilitate the planning of a one-semester course: these volumes aim at being their main teaching tool. More specifically, each part identifies an independent teaching module, while each chapter corresponds to about two weeks of lecturing.

I cannot conclude this general introduction without thanking the many students who, over the years, have attended my courses in condensed matter physics at the University of Cagliari. Through the continuous exchange of ideas with them I have gradually understood how best to organize my teaching and the corresponding study material. As a matter of fact, the contents that I have collected in these volumes were born from this very fruitful dialogue.

LUCIANO COLOMBO
Department of Physics
University of Cagliari
Cagliari, June 2019

Introduction to the first edition

Atoms are the basic building blocks of any material system: their physics rules over the behaviour of basically everything. Therefore, even if we limit ourselves to just considering what is found on our planet, the atomistic picture is really needed to understand at the most fundamental level the physical behaviour of organic and inorganic matter, biological molecules and systems, as well as geological materials.

Accordingly, some substantial basic notion on atomic physics must be recognised as inalienable from a modern undergraduate course in physics, as well as in materials science, chemistry, and, likely, engineering too. Since the first step in atomic aggregation consists in the formation of molecules, any tutorial introduction to the topic must necessarily also contain an introduction to molecular physics.

This state of affairs defines the main goal of this volume, namely: to present a modern and unified tutorial account of the fundamental physics of atoms and molecules. A modern approach does require the use of quantum mechanics, while a unified perspective dictates looking at a molecule as a bound system of atoms, still retaining some of their properties. Both features are fully exploited in this volume.

This Primer is divided in three parts: the first one provides a brief historical introduction on early atomic physics and outlines the kernel of non-relativistic quantum mechanics, the second one is addressed to atomic physics, and the third one deals with molecules. Eight appendices are added to the text, each focussed on some technical development which, at first reading, can be skipped without compromising the general understanding of the arguments developed in the main text. A bibliography is added to each chapter as a guideline for further reading.

The volume contains many figures, most of which are 'conceptual', i.e., they are basically intended to provide a graphical representation of the main ideas and concepts developed in the written part. Tables with numerical values of important physical properties are included as well, in the attempt to provide the reader with information useful to 'quantify' the physical results presented. Finally, a list of all the mathematical symbols used in the volume is given at the beginning as an orientation guide while reading.

LUCIANO COLOMBO
Department of Physics
University of Cagliari
Cagliari, June 2019

About the author

Luciano Colombo

Luciano Colombo received his doctoral degree in physics from the University of Pavia (I) in 1989 and then he was a post-doc at the École polytechnique fédérale de Lausanne (CH) and at the International School for Advanced Studies (I). He became assistant professor (tenured) at the University of Milano (I) in 1990, next moving to the University of Milano-Bicocca (I) in 1996 for an equivalent position. In 1999 he was appointed associated professor at the University of Cagliari (I) where in 2002 he became full professor of theoretical condensed matter physics. Since 2015 he has been fellow of the 'Istituto Lombardo—Accademia di Scienze e Lettere' (Milano, I). In 2021 he has been appointed as Vice-Rector for Research of the University of Cagliari (I) and 'Managing Editor' of the 'Materials Physics' section of The *European Physical Journal—Plus*. He has been the principal investigator of several research projects addressed to solid-state and materials physics problems, the supervisor of more than 80 students (at bachelor, master, and PhD level), and the mentor of about 20 post-docs. He is the author, or coauthor, of more than 297 scientific articles[1] and 10 books (this included). More about him can be found at: http://people.unica.it/lucianocolombo.

---

[1] Source: Scopus—more bibliometric information can be found at: https://www.scopus.com/results/authorNamesList.uri?st1=Colombo&st2=Luciano&institute=Cagliari&origin=searchauthorlookup.

List of symbols

α	fine structure constant
$\alpha(\omega)$	absorption coefficient (Beer law)
α_{elec}	electric dipole polarizability
α_{hf}	hyperfine splitting constant
Γ	loss factor of a LASER device
ε_0	vacuum permittivity
ζ	atomic valence
μ_B	electron Bohr magneton
μ_N	nuclear magneton
μ_0	vacuum permeability
$\xi(r)$	spin–orbit coupling constant (hydrogenic atoms)
ξ_{LL}	orbit–orbit coupling constant (vector model of the atom)
ξ_{LS}	spin–orbit coupling constant (vector model of the atom)
ξ_{SS}	spin–spin coupling constant (vector model of the atom)
$\phi_{nlm}(\mathbf{r})$	Slater-type orbital
$\varphi(\mathbf{r})$	single-electron spin-orbital
χ_A	antisymmetric total spin wavefunction of an electron system
χ_S	symmetric total spin wavefunction of an electron system
$\psi_{ab}(\mathbf{r})$	anti-bonding molecular orbital
$\psi_b(\mathbf{r})$	bonding molecular orbital
$\psi_{nlm_l}(\mathbf{r})$	single-electron hydrogenic wavefunction
$\Psi(\mathbf{r}_1, \mathbf{r}_2,...)$	generic total space wavefunction of an electron system
Ψ_A	antisymmetric total space wavefunction of an electron system
Ψ_S	symmetric total space wavefunction of an electron system
ω_0	fundamental oscillation frequency of a diatomic molecule (harmonic approximation)
ω_{Larmor}	Larmor precession frequency
ω_{Thomas}	Thomas precession frequency
a_0	Bohr radius
A	mass number
$\mathbf{A}$	vector potential for the electromagnetic field
$A_{21}^{(\nu,\omega)}$	Einstein coefficient for spontaneous emission in the ν- or ω-representation
B	rotational constant in a diatomic molecule
$B_{12}^{(\nu,\omega)}$	Einstein coefficient for stimulated absorption in the ν- or ω-representation
$B_{21}^{(\nu,\omega)}$	Einstein coefficient for stimulated emission in the ν- or ω-representation
c	speed of light
d_{elec}	induced electric dipole moment
e	electron charge (absolute value)
E_{ab}	molecular energy in an anti-bonding state
E_b	molecular energy in a bonding state
E_{diss}	dissociation energy of a diatomic molecule
$E_e^{(\mathbf{R})}$	electronic energy of a molecule in the nuclear configuration $\mathbf{R}$
E_{GS}	ground state energy (both in atomic and molecular systems)
E_n	nuclear (vibrational+rotational) energy of a molecule
E_{so}	spin–orbit interaction energy
$E_p(R)$	potential energy of a diatomic molecule (R is the internuclear distance)
E_{rot}	rotational energy of a molecule

E_{vib}	vibrational energy of a molecule
E_{T}	total energy of a molecule
F	Faraday constant
g_j	Landé g-factor
g_L	orbital g-factor
g_N	nuclear g-factor
g_S	spin g-factor
$G(\omega)$	gain factor of a LASER device
h	Planck constant
$\hbar$	reduced Planck constant $\hbar = h/2\pi$
H_{AA}	Coulomb integral in a diatomic molecule
H_{AB}	resonance integral in a diatomic molecule
I	moment of inertia of a molecule
$I_\alpha^{1\text{b}}$	single-electron Hartree–Fock integral
$I_{\alpha\beta}^{2\text{b,d}}$	two-electron Hartree–Fock integral (direct term)
$I_{\alpha\beta}^{2\text{b,ex}}$	two-electron Hartree–Fock integral (exchange term)
$I_{nl,n'l'}$	Coulomb integral calculated for the pair of nl and $n'l'$ hydrogenic states
j	total angular momentum quantum number (single electron)
j_{tot}	total angular momentum quantum number for the full set of electrons
$\mathbf{J}$	total (orbital+spin) angular momentum vector (single electron)
$\mathbf{J}_{\text{tot}}$	total (orbital+spin) angular momentum vector for the full set of electrons
$\mathbf{k}$	wavevector of an electromagnetic wave
k_{B}	Boltzmann constant
$K_{nl,n'l'}$	exchange integral calculated for the pair of nl and $n'l'$ hydrogenic states
l	orbital angular momentum quantum number (single electron)
l_{tot}	total orbital angular momentum quantum number for the full set of electrons
$\mathbf{L}$	orbital angular momentum vector (single electron)
L_z	z-component of the orbital angular momentum vector (single electron)
$\mathbf{L}_{\text{tot}}$	total orbital angular momentum vector for the full set of electrons
m_l	orbital magnetic quantum number (single electron)
$m_{l_{\text{tot}}}$	total orbital magnetic quantum number for the full set of electrons
m_{e}	electron mass
m_{p}	proton mass
m_s	spin magnetic quantum number (single electron)
$m_{s_{\text{tot}}}$	total spin magnetic quantum number for the full set of electrons
$\mathbf{M}_L$	orbital magnetic moment vector (single electron)
$\mathbf{M}_{L_{\text{tot}}}$	total orbital magnetic moment vector for the full set of electrons
$\mathbf{M}_N$	nuclear magnetic moment vector of an atom
$\mathbf{M}_S$	spin (or intrinsic) magnetic moment vector (single electron)
$\mathbf{M}_{S_{\text{tot}}}$	total spin (or intrinsic) magnetic moment vector for the full set of electrons
$\mathbf{M}_{J_{\text{tot}}}$	total (orbital+spin) magnetic moment vector of an atom (electron contribution)
$n^{2S+1}L_J$	spectroscopic symbol for atomic levels (with $n = 1, 2, 3,\ldots$ and $L = S, P, D,\ldots$)
n_{d}	degree of degeneracy of an hydrogen quantum state
$\mathbf{N}$	total nuclear angular momentum vector of an atom
$\mathcal{N}_{\text{A}}$	Avogadro number
$P_{nl}(r)$	radial distribution function (or radial probability)
$\mathcal{P}_{1\to 2}$	probability per unit time for a quantum transition from state '1' to state '2'
$R_{nl}(r)$	hydrogen radial wavefunction

$\bar{R}_{nl}(r)$	radial central-field wavefunction in a multi-electron atom
$\mathcal{R}_{\mathrm{H}}$	Rydberg constant for the hydrogen atom (infinite nuclear mass approximation)
$\bar{\mathcal{R}}_{\mathrm{H}}$	Rydberg constant for the hydrogen atom with finite nuclear mass
s	spin quantum number (single electron)
s_{tot}	total spin quantum number for the full set of electrons
$\mathbf{S}$	spin angular momentum vector (single electron)
S_z	z-component of the spin angular momentum vector (single electron)
S_{AB}	overlap integral in a diatomic molecule
$\mathbf{S}_{\mathrm{tot}}$	total spin angular momentum vector for the full set of electrons
u_ν^{BB}	spectral density of the blackbody radiation
V_{cf}	central-field potential
V_{ee}	electron–electron Coulomb interaction potential
V_{ne}	nucleus–electron Coulomb interaction potential
V_{nn}	nucleus–nucleus Coulomb interaction potential
V_{nucl}	nuclear potential energy in a diatomic molecule (including the adiabatic term)
V_{TF}	Thomas–Fermi potential
w	atomic weight
W_α	ionisation work (electron accommodated on the φ_α spin-orbital)
$Y_{lm_l}(\theta, \phi)$	spherical harmonic function
Z	atomic number

Fundamental physical constants

Physical constant	Symbol	Value[+]	Units
Atomic mass unit	a.m.u.	$1.660\,539 \times 10^{-27}$	kg
Avogadro number	N_A	$6.022\,140 \times 10^{23}$	mol^{-1}
Bohr magneton	μ_B	$9.274\,010 \times 10^{-24}$	$J\,T^{-1}$
Bohr radius	a_0	$5.291\,772 \times 10^{-11}$	m
Boltzmann constant	k_B	$1.380\,649 \times 10^{-23}$	$J\,K^{-1}$
Electron charge	e	$1.602\,176 \times 10^{-19}$	C
Electron rest mass	m_e	$9.109\,383 \times 10^{-31}$	kg
Electron charge-to-mass ratio	e/m_e	$1.758\,820 \times 10^{-11}$	$C\,kg^{-1}$
Faraday constant	F	$9.648\,533 \times 10^{4}$	$C\,mol^{-1}$
Fine structure constant	α	$1/137$	
Neutron rest mass	m_n	$1.674\,927 \times 10^{-27}$	kg
Planck constant	h	$6.626\,070 \times 10^{-34}$	J s
Planck constant (normalised)	$\hbar$	$1.054\,571 \times 10^{-34}$	J s
Proton rest mass	m_p	$1.672\,621 \times 10^{-27}$	kg
Speed of light (vacuum)	c	$2.997\,924 \times 10^{8}$	$m\,s^{-1}$
Stefan constant	σ_S	$5.670\,374 \times 10^{-8}$	$W\,m^{-2}\,K^{-4}$
Molar gas constant	R	$8.314\,462$	$J\,K^{-1}\,mol^{-1}$
Vacuum magnetic permeability	μ_0	$1.256\,637 \times 10^{-6}$	$N\,A^{-2}$
Vacuum electric permittivity	ε_0	$8.854\,187 \times 10^{-12}$	$F\,m^{-1}$

[+] The reported values of fundamental physical constants are published and recommended for international use by the CODATA Task Group.
See https://www.nist.gov/pml/fundamental-physical-constants website for more information.

Part I

Preliminary concepts

IOP Publishing

Atomic and Molecular Physics (Second Edition)
A primer
Luciano Colombo

Chapter 1

The overall picture

Syllabus—A brief historical and conceptual review is presented on early atomic physics, suggesting that matter is discrete, physical laws are quantum, and all phenomena comply with matter–wave duality. It is also argued that quantum physics is inherently probabilistic. These achievements represent as a whole a conceptual exit from the realm of classical physics which, in turn, calls for a novel formal theory embodying all such features.

1.1 The atomistic structure of matter

1.1.1 Do atoms exist?

Matter offers to our senses as a continuum, categorized as a solid, liquid or gas by simply observing whether it has, respectively, a definite volume and shape, or just a definite volume, or none of them. According to this continuum picture, we can split (in the case of solids) or separate (in the case of liquids and gases) any given matter sample into two (or more) parts and, in principle, we can repeat this operation by an arbitrary number of times, still obtaining continuum matter specimens (eventually very small).

This macroscopic description is contrasted by a more accurate observation of the structure of matter, as first elaborated at the dawn of the XIXth century through the combined efforts of a number of scientists[1]. On the basis of chemical evidence, it was observed that matter can be formed either by pure substances (or *elements*) and by *compounds* mixing two or more elements. Next, a law of multiple definite proportions for any compound was formulated: *the fixed amount of a given element and the corresponding amounts of any other element needed to form a compound are in the ratio of small integer numbers.* For instance: for any given 'quantity' of, say, oxygen there are needed two 'quantities' of hydrogen to form that compound we

---

[1] Mainly: J L Proust in 1801, J Dalton in 1807–8, J L Gay-Lussac in 1801, and A Avogadro in 1811. A comprehensive historical perspective is reported elsewhere [1, 2].

name water. These empirical observations suggest that matter in fact cannot be arbitrarily divided into increasingly smaller parts, still maintaining its original chemical properties. Rather, a new picture was emerging in that: matter in whatever state of aggregation is made by elements or compounds; each element is made by elementary constituents, hereafter referred to as *atoms*; and compounds are formed by combining numbers of different atoms in simple ratios. It was a natural guess assuming that atoms of the same element are identical in nature and have just the same weight. Since the atomistic structure of matter cannot be addressed by our senses, it was further assumed that atoms are really very small.

Once the concept of the atom is accepted, we can elaborate a picture of matter rather different from the continuum one, namely: any solid of fluid substance is formed either by an assembly of atoms of just one kind or by an assembly of *molecules*. These latter material entities are the smallest parts of any compound substance that determine its chemical properties. It is then straightforward to introduce the concept of *mole* as the most natural unit for quantifying the actual amount of matter forming a given specimen: a mole is that quantity weighting in grams the same number expressing the atomic or molecular weight[2]. Based on thermodynamical arguments, Avogadro argued that *a mole of any substance contains the very same number N_A of atoms or molecules*, whether the substance is, respectively, elemental or compound. Such a number was determined by several different experimental methods [2] based on thermodynamical, electrochemical, or x-ray diffraction measurements; its more accurate value was set in 2019 as

$$N_A = 6.022\,140 \times 10^{23} \text{ mol}^{-1} \tag{1.1}$$

which allows for a qualitative estimation of the typical size of an atom. By assuming that a solid matter sample with uniform mass density ρ is made by close-packed spherical atoms[3] with diameter d, we simply obtain an order-of-magnitude estimation of the typical *atom size*: $d \sim 1 - 2 \times 10^{-10}$ m.

A robust experimental confirmation of the atomistic hypothesis is provided by the *kinetic theory of gases*, as originally cast by J C Maxwell and R J E Clausius: under the sole guess that an ideal gas is made by non-interacting, point-like, and massive atoms and by calculating their mechanical actions on the walls of the container, it is in fact possible to work out the well-known state equation $PV = nRT$, where P, V, and n are, respectively, the pressure, volume, and number of moles of the gas at the equilibrium temperature T (R is the molar gas constant). In this way, not only does the atomistic kinetic theory provide predictions about the thermodynamical behaviour of a gas directly confirmed by experiments, but also the phenomenological concept

---

[2] The atomic weight is the average of the mass values of a given element, weighted by the natural abundance of isotopes of that atom. Mass values are typically given in atomic mass units (a.m.u.), corresponding to 1/12 of the mass of the ^{12}C isotope. It is set 1 a. m. u. $=1.660\,539 \times 10^{-27}$ kg. The concept of isotope will be discussed later in this chapter.

[3] The guess about the spherical shape of an atom was originally elaborated by Dalton as a byproduct of the law of definite proportions.

of temperature is naturally linked to the thermal motion of the elementary constituents of matter: this is indeed a major conceptual achievement[4].

However, a direct observation of single atoms is not possible by ordinary optical microscopy since visible light has a wavelength 400 nm $\leqslant \lambda \leqslant$ 700 nm, much larger than the typical atom size. Optical techniques can nevertheless provide another indirect proof of the existence of atoms, like in the noteworthy case of the *Brownian motion* firstly observed by R Brown in 1827: small particles suspended in a liquid move randomly, generating irregular trajectories. The physical origin of such irregular paths is due to a sequence of scattering events occurring among the suspended particles and the thermally activated atoms forming the liquid. When a particle is hit by a fast moving atom, its trajectory is suddenly deflected. Since the atomic thermal motion is non-directional, multiple scattering events generate an irregular trajectory which can in fact be observed by means of an optical microscope. The theory of random Brownian motion was originally developed by A Einstein in 1905 who proved its close relation to atomic diffusion [6].

X-ray scattering by solids provides an alternative strong support to the atomistic picture. By collecting the diffracted x-ray beams emerging from a crystalline solid sample, we obtain a regular pattern (diffractogram) consisting in high-intensity diffraction peaks separated by low-intensity regions. M von Laue explained such a result in 1912 by assuming that the solid was formed by a periodic distribution of atoms, each acting as a scattering centre of the incoming beam. Diffraction is, therefore, a collective event of absorption and re-emission of electromagnetic waves by all the atoms. By analysing the regular pattern reported in a typical diffractogram, it is possibile to map the actual distribution of atoms in space: such a solid will be named crystal, to emphasize its periodic atomic architecture.

A detailed description of the remaining large variety of experiments addressing the atomistic structure of matter falls beyond the scope of this short historical survey. However, it is worth remarking that the ultimate proof about the existence of atoms is provided by modern nano-scopies, a remarkable payoff indeed for the atomistic hypothesis these nanotechnologies are based on. In scanning electron microscopy (SEM) an electron[5] beam is scanned on the surface of a sample, emitting secondary electrons and x-rays. By analysing such emissions it is possible to elaborate a map of the surface atoms with a typical resolution in the range 1–20 nm. In transmission electron microscopy (TEM) a high-energy electron beam is transmitted through a thin sample in consequence of multiple scattering events occurring among the impinging electrons and the atoms forming the thin sample. Once again, it is possible to elaborate a picture of the very atomic structure of the sample by analysing the diffracted paths with a remarkable precision down to the 0.1 nm scale. A comprehensive description of electron microscopy techniques is found elsewhere [7]. The surface atomistic structure can alternatively be mapped by

---

[4] This remarkable result paves the way for a *microscopic* formulation of thermodynamics (which, in its most advanced setup, is known as statistical mechanics [3–5]).

[5] The true existence of electrons, namely elementary negatively-charged particles, is discussed later in this chapter.

an atomic force microscope [8]: here the bending of a sharp cantilever tip scanning the surface is measured, extracting information about the surface roughness due to the actual distribution of atoms.

In conclusion, the query 'Do atoms exist?' has a clean and experimentally well-supported answer: yes, atoms do exist. Condensed matter is in fact an assembly of interacting atoms, the description of whose physical properties will be the main topic of this volume. The paramount importance of the *atomistic hypothesis* is marvellously summarized in the following statement by R P Feynman: '*If, in some cataclysm, all of scientific knowledge were to be destroyed, and only one sentence passed on to the next generation of creatures, what statement would contain the most information in the fewest words? I believe it is the atomic hypothesis (or the atomic fact, or whatever you wish to call it) that all things are made of atoms, little particles that move around in perpetual motion, attracting each other when they are a little distance apart, but repelling upon being squeezed into one another. In that one sentence, you will see, there is an enormous amount of information about the world, if just a little imagination and thinking are applied.*' [9].

1.1.2 What are atoms made of?

None of the experimental evidences discussed in the previous section provides information about the inner structure of an atom, nor the phenomenological theories based on the early atomistic hypothesis indeed require any knowledge about this issue (for instance, the kinetic theory of gases assumes structureless atoms).

There are, however, direct and indirect evidences that matter—made of atoms—does contain electrically charged particles. To name just a few: (i) by applying an electric field to a polar liquid, an electrolytic current is observed and explained, as originally proposed by M Faraday, in terms of dissociation of molecules into positive and negative constituents, hereafter referred to as *ions*, drifting in opposite directions; (ii) particles like α or β ones (which we nowadays recognize as helium nuclei and electrons, respectively) emitted by radioactive substances are differently deflected by an external magnetic field due to the Lorentz force, thus proving that they carry a charge; (iii) a charge current is observed in metals under bias, as proved by electrical measurements. We must eventually conclude not only that matter is made by atoms, but also that *atoms are made by substructures with either positive or negative electric charge*. The differently charged constituents of atoms have different masses as well.

Chemistry as a whole supplies evidence[6] that the *negative substructures are electrons*, i.e., particles carrying just one negative *elementary electric charge* $e = 1.602176 \times 10^{-19}$ C with a mass $m_e = 9.109\,383 \times 10^{-31}$ kg $= 5.485\,799 \times 10^{-4}$ a.m.u. The first accurate estimation of the e/m_e ratio was provided by J J Thomson in 1897, after various attempts. The actual value of the electron mass was later measured by the suspended oil droplet experiment performed by R A Millikan in

---

[6] Through, e.g., the concept of atomic valence ζ, the elementary theory of ionic bonding, a variety of electrochemical phenomena, and the reduction–oxidation reactions where the oxidation state of an atom is changed [10]. We remark that the negative sign is assigned to the electron charge by convention, supported by Hall effect measurements [9].

1909–10. Previously, the concept of elementary electron (that is, the very fact that electric charge is not infinitely divisible) was recognized by M Faraday in 1883 when he worked out the fundamental law of electrolysis in the form

$$M = \frac{1}{F}\frac{w}{\zeta}Q \tag{1.2}$$

stating that the mass M of a substance liberated at an electrode over a given time is always proportional to the net charge Q passed through the electrolytic solution in the same time. In equation (1.2) ζ and w are, respectively, the valence of the atoms and the molar mass of the liberated substance, while $F = 9.648\,533 \times 10^4$ C mole^{-1} is the so-called Faraday constant. Since a charge $Q = 96485.3$ C liberates exactly 1 mole of a monovalent ($\zeta = 1$) substance, the elementary electric charge is naturally defined as $e = F/\mathcal{N}_A$, which provides the value of the electron charge reported above.

The true existence of electrons is confirmed in many ways [2], including: the photoelectric effect (i.e., the emission of electrons by a metal surface illuminated by ultraviolet radiation), the deflection of cathode rays by a magnetic field or, finally, the thermionic emission phenomenon (i.e., the emission of electrons from solids heated to a suitably high temperature).

Since atoms in their normal state are electrically neutral (likewise uncharged condensed matter), we must conclude that *each atom is an electrically complex system, containing an equal amount of negative and positive charge*. While the former is provided by a given number of electrons, the latter is in a still unknown form. In brief: if an atom contains Z electrons, it must also contain other constituents with a total positive charge $+Ze$. The number Z is known as *atomic number*. A further observation consists in comparing m_e to the mass m_H of the lightest atom, namely hydrogen (H), containing just one electron: it results as $m_H \gg m_e$, thus implying that *most of the atomic mass must be associated with the positively charged constituents*.

1.1.3 The nuclear atom

An obvious next question immediately arises: how are negative and positive charges distributed over the volume of the atom? In an attempt to reply, we focus on hydrogen for the sake of simplicity and adopt the atomic model firstly introduced by J J Thomson in 1904 (also known as 'plum pudding model'): the sole electron of the H atom is located within a continuous distribution of positive charge for which a spherical volume of radius r_H is assumed. It is a straightforward exercise of electrostatics to prove that an electron so embedded undergoes harmonic oscillations around the centre of the sphere with a frequency

$$\nu = \frac{1}{2\pi}\sqrt{\frac{\rho e}{3m_e\epsilon_0}} \tag{1.3}$$

where ϵ_0 is the vacuum dielectric permittivity and $\rho = 3e/4\pi r_H^3$ is the uniform density of the positive charge distribution. Since the electron is charged, such oscillations generate electromagnetic (e.m.) radiation at the same frequency. According to the

Thomson model, therefore, a hydrogen atom would emit e.m. waves at the sole frequency predicted by equation (1.3). Unfortunately, this is qualitatively wrong since *the experimentally observed emission spectrum of a hydrogen atom consists in a multiple sequence of discrete emission lines* at very many different frequencies.

The compelling proof that the Thomson model is wrong was worked out in 1911 by E Rutherford. A beam of positively charged α particles emitted by a radioactive source was made incident upon a gold metallic foil, thin enough (~ 1 μm) to allow α particles to pass through. Transmitted particles were then collected and their trajectories accurately identified, resulting in three kinds: (i) most particles were only slightly deflected; (ii) an appreciably smaller number of particles was deflected at comparatively much larger angles; and (iii) an even smaller number of particles was in fact backscattered. By considering that deviation angles were in any case $\geqslant 1°$, it was possible to exclude that α particles were scattered by the electrons contained in the foil[7]. It was as well shown that the collected trajectories are incompatible with the Thomson model, according to which no large deflection angles would be possible (this is the second major failure of such a model). The novel concept introduced by Rutherford to overcome this impasse consisted in assuming that *all the positive charge of an atom is concentrated in a very small region*, rather than in the whole atomic volume. The small region is named *nucleus* and, following our previous conclusion about the mass distribution, it also contains most of the atomic mass. Rutherford derived an expression for the deflection of a beam of α particles by point-like positive charges (the atomic nuclei) using classical scattering theory [11], which proved to be very effective in predicting the observed trajectories.

The Rutherford model is commonly referred to as the *nuclear atomic model*, putting emphasis on the presence of the nucleus, or as the *planetary atomic model*, since electrons—similarly to planets gravitating around a star—are described as moving around a massive point-like nucleus under the action of its Coulomb central field.

An important refinement of the Rutherford planetary model consists in observing that the atomic nucleus is in fact neither point-like nor structureless, as supported by a huge body of experimental evidence [12]. The first conclusion is for instance motivated by the departure of the cross-section from the Rutherford formula for head-on collisions, i.e., when the kinetic energy of the α particle is large enough to make the approach distance to an atom so small as to be comparable to the actual size of the nucleus $\sim 10^{-14}$ m. Nuclear physics shows that a nucleus contains both positively charged *protons* and uncharged *neutrons* [12]. For our purposes it is sufficient to treat protons and neutrons as point-like particles with the same mass.

The chemical properties of an atom are determined by its atomic number Z, i.e., by the number of electrons or, equivalently, by the nuclear charge $+Ze$. However, it is observed that atoms exist with different nuclear mass but the same chemistry. This implies that for a given Z there can exist several nuclei differing in the number N of neutrons: they are named *isotopes* and recognized according to their *mass number* $A = Z + N$ [12]. On the other hand, by removing electrons from an atom, we obtain

---

[7] This conclusion relies on scattering theory which predicts deflection angles of $\sim 0.1°$ if the mass of the scatterer and of the scattered particles correspond to that of an electron and of an α particle, respectively.

a positively charged object, i.e., an *ion*. If, upon multiple removal, just one single electron is left we obtain a very peculiar atomic configuration which is addressed as *hydrogenic atom* (or hydrogen-like atom), with reference to the most simple chemical element *hydrogen* characterized by $Z = 1$ and a nucleus formed by just one proton. The electronic structure of all hydrogenic atoms can be treated by the same formal approach used for H.

1.2 The quantum nature of physical laws

1.2.1 Atomic spectra: failure of classical physics

The above achievements about the structure of matter clearly bring out the next objective, namely: understanding the inner physics of an interacting planetary system of one or more electron(s) and a nucleus.

Let us consider the hydrogen atom and assume that for its ground state the radius of the circular electronic orbit is $a_0 = 0.5291772$ Å, hereafter referred to as *Bohr radius*[8]. In this classical model, the electron is rotating at speed

$$v_0 = \sqrt{\frac{e^2}{4\pi\epsilon_0 m_e a_0}} \sim 2 \times 10^6 \text{ m s}^{-1} \tag{1.4}$$

feeling a linear acceleration

$$a = \frac{v_0^2}{a_0} \sim 9 \times 10^{22} \text{ m s}^{-2} \tag{1.5}$$

since it moves under the action of the Coulomb central field generated by the nucleus. Now, classical physics dictates [9] that any accelerated particle of charge $\pm e$ irradiates electromagnetic energy at power

$$P = \frac{e^2 a^2}{6\pi\epsilon_0 c^3} \tag{1.6}$$

where c is the speed of light. In the atomic case we are discussing, this corresponds to $P_H \sim 4 \times 10^{11}$ eV s^{-1} representing the emission power of hydrogen[9]. Chemistry provides evidence that the ionization energy of the H atom in its ground state is 13.6 eV, that we can consider as the net amount of energy stored in a bound electron–proton planetary system. Since such an energy is dissipated by irradiation at the rate P_H, we immediately obtain that a hydrogen atom would lose all its energy in about 10^{-11} s. Over such an astonishingly short lapse of time the electron speed would vanish, likewise the orbital radius: the electron 'falls' on the nucleus. By generalizing this result we come to a rather disturbing conclusion, namely: according to classical physics matter should not be stable or, equivalently, we should not observe the Universe in its present

---

[8] By setting the actual value of a_0 we have anticipated a fundamental result of early atomic physics which will be derived later on in section 1.2.3.

[9] As customary in atomic physics we measure energies in units of eV, corresponding to the kinetic energy of an electron accelerated by an electrostatic potential difference of 1 volt.

appearance. Bad enough, this discouraging result is not the only failure of the classical planetary atomic model: the trajectory of the electron falling motion is predicted to be a spiral whose radius continuously varies as $r(t) = a_0 - \mathcal{A}t^{1/3}$, where $\mathcal{A}$ is a suitable constant. This in turn implies that the emission spectrum of the hydrogen atom should be continuous, completely at odds with experimental evidence as already commented in section 1.1.3.

The conclusion we can draw is unequivocal: *the Rutherford model cannot be adopted in bundle with the laws of classical physics*. Since such a model has proved to be very successful in describing α particle scattering, it is really hard to discard it: in fact, this was not the option chosen by scientists working in this field. Rather, some new physics is needed which must be inherently non-classical.

1.2.2 A necessary digression

In searching for some non-classical concepts possibly useful to our advancement, we will briefly discuss the physics of blackbody radiation and the photoelectric effect. A more extensive treatment of both phenomena falls beyond the scope of this book and can be found elsewhere [2, 13]. The emerging new concepts will be then applied to the interpretation of the atomic spectra in the next section.

1.2.2.1 Blackbody radiation
It is known that any solid body can both absorb and emit light. These phenomena are typically described by the *spectral absorption* a_ν or *spectral emission* e_ν coefficients, respectively defined as the e.m. power absorbed or emitted by the unit of surface area of the body at frequency ν. While a_ν and e_ν individually depend on the actual physico-chemical properties of the solid and the morphological features of its surface, their ratio is given, at any frequency, by a universal function of the sole temperature T, as first discussed by G Kirchhoff in 1859–62. A blackbody is an ideal system characterized by $a_\nu = 1$. In practice, it can be mimicked by a solid sample containing a cavity with blackened inner surfaces and a small hole through which e.m. radiation can be exchanged with the surrounding ambient: once a radiation penetrates into the cavity, it can very unlikely escape (absorption close to 100%) because of the small size of the orifice; on the other hand, thermal radiation can be emitted through the very same aperture (blackbody radiation).

By bringing such a system to an equilibrium temperature T, it is possible to record its emission spectrum, customarily addressed as *the blackbody radiation*: it is a non-monotonic function of the e.m. frequency, which also depends on temperature. The experimental findings are summarized by two laws, respectively: (i) in 1879 J Stefan derived that the total e.m. power $P_{tot}(T)/A$ emitted per unit area is given by

$$\frac{P_{tot}(T)}{A} = \sigma_S T^4 \tag{1.7}$$

where A is the area of the emitting surface and $\sigma_S = 5.670\ 374 \times 10^{-8}$ W m^{-2} K^{-4} is nowadays known as *Stefan constant*; (ii) in 1893 W Wien proved that if ν_{max} is defined as the frequency at which we observe the maximum emission intensity, than it holds that

$$\frac{\nu_{\max}}{T} = \text{constant} \tag{1.8}$$

a result which is usually referred to as the *Wien displacement law*.

Attempts based on classical physics turned out to be unsuccessful in elaborating a formula for the blackbody radiation accounting for both equations (1.7) and (1.8). In particular, at the beginning of the XXth century J W Strutt (Lord Rayleigh) and J Jeans independently calculated the *spectral energy density*[10] u_ν of the trapped e.m. wave as

$$u_\nu = n_\nu \langle E_\nu \rangle \tag{1.9}$$

where n_ν is the spectral density of e.m. modes trapped in the cavity[11] and $\langle E_\nu \rangle$ is the mean energy of the mode with frequency ν. This latter quantity was guessed according to the classical equipartition theorem [4]

$$\langle E_\nu \rangle = k_B T \tag{1.10}$$

where $k_B = 1.380\,649 \times 10^{-23}$ J K^{-1} is the Boltzmann constant. This guess can be reconciled with the atomistic picture we are developing by observing that each vibrating atom in the cavity wall can be looked at as a radiator emitting an e.m. wave at the same frequency it oscillates. The energy distribution of the atomic radiators and the e.m. field in the cavity must be the same since we have set an equilibrium condition. If not, a net energy flux should occur between matter and radiation. Finally, the mean energy of classical oscillators is provided by the equipartition theorem. In brief, it has been assumed the same classical mean energy for atomic radiators and e.m. modes with identical frequency. On the other hand, the explicit calculation of

$$n_\nu = \frac{8\pi}{c^3} \nu^2 \tag{1.11}$$

was obtained through a somewhat more challenging calculation[12].

---

[10] Or, equivalently, the volume energy density of the e.m. radiation field at frequency ν.

[11] Or, equivalently, the number of e.m. standing waves with frequency ν trapped into the unit volume of the blackbody cavity.

[12] There is a fully classical way to calculate this quantity, directly counting the number of allowed standing waves trapped in a cavity, under the constraint that they must have nodes at any end of the cavity [13]. However, such a classical calculation is a bit tricky and a much simpler way to proceed consists in using some quantum concepts. Let us consider at first a free particle confined in a volume V. It is a well-known result of elementary quantum mechanics [14–16] that the quantity $(1/2\pi^2\hbar^3)p^2 dp$ represents the number density of allowed quantum states with corresponding momentum in the interval $[p, p + dp]$. Next, by profiting from a result discussed in detail in section 1.4.1, we remark that each blackbody radiation mode of frequency ν consists of free pseudo-particles named photons with energy $E^{\text{photon}} = h\nu$. Their momentum $p = h/\lambda$ is provided by the de Broglie relation, where $\lambda = c/\nu$ is the mode (or photon) wavelength. By replacing these relations in the above expression for the number of allowed states, we easily obtain equation (1.11).

By inserting equations (1.11) and (1.10) into equation (1.9) we get the *classical Rayleigh–Jeans formula*

$$u_\nu = \frac{8\pi}{c^3} \, k_B T \, \nu^2 \tag{1.12}$$

for the spectral density emitted by a blackbody equilibrated at temperature T. As anticipated, this formula is wrong: not only is it inconsistent with both Stefan and Wien laws but, more importantly, it provides an infinite value for the total energy emitted by the blackbody, as a consequence of the $\sim\nu^2$ divergence. This was called ultraviolet catastrophe.

The impasse was later solved by M Planck in 1900–4 by introducing a novel concept with no analogue in classical physics: he assumed that *the energy spectrum of the radiation trapped within the blackbody cavity was discrete*, instead of continuous. More specifically, by treating each normal mode of such e.m. radiation as a harmonic oscillator (h.o.), it was imposed that its energy can only take *discrete values* given by

$$E_n^{\text{h.o.}} = nh\nu \quad \text{with} \quad n = 0, 1, 2, 3,\ldots \tag{1.13}$$

where h is a suitable constant with the proper units to convert a frequency into an energy. Under this hypothesis, Planck derived for the energy density of the blackbody radiation u_ν^{BB} the renowned equation

$$u_\nu^{BB} = \frac{8\pi h}{c^3} \, \frac{\nu^3}{e^{h\nu/k_B T} - 1} \tag{1.14}$$

which turned out to be in spectacularly good agreement with experimental data, provided that the following fitted value

$$h = 6.626\,070 \times 10^{-34} \text{ J s} \tag{1.15}$$

is used: hereafter h will be named *Planck constant*.

While equation (1.14) reduces to the Rayleigh–Jeans equations in the limit $\nu \to 0$ and shows no ultraviolet catastrophe, it is also nicely consistent with equations (1.7) and (1.8). We nevertheless remark that this success required a new odd concept, namely: the *quantization of the energy spectrum of a harmonic oscillator*. This concept has been forced into the theory with actually no other justification than heuristic[13].

---

[13] We also remark that the original derivation of the Planck equation is no longer considered as physically sound since it was based on two wrong assumptions, namely: (i) the discretized energy of the harmonic oscillators was missing the 'zero point energy' contribution predicted by rigorous quantum mechanics and (ii) the classical Boltzmann statistics was used for them. In this respect, the original derivation is fortuitous. Nevertheless, the original Planck argument was seminal in assuming that the interaction between electromagnetic radiation and matter occurs through emission/absorption of discrete energy quanta $h\nu$ and, therefore, in this respect the Planck equation represents one of the major achievements of early quantum mechanics.

1.2.2.2 Photoelectric effect

Although inconsistent with any previous knowledge, the quantization procedure was soon used by A Einstein in 1905 to explain the puzzling photoelectric effect discovered by H Hertz in 1887. In this case, the electric current emerging from a metal cathode is collected by the anode plate, under light illumination. It was observed—with no explanation based on classical physics—that: (i) the current is only observed if the incoming monochromatic e.m. wave has frequency in the ultraviolet region[14]; (ii) in this condition, the current intensity is nonzero even for zero bias and it steadily grows by applying a raising positive voltage, until reaching a maximum saturation value; at variance, (iii) under increasing negative voltage the observed current reduces almost linearly, eventually reaching a zero value for a suitable stopping voltage. Interestingly enough, by repeating the experiment with the same apparatus and radiation, but only increasing its intensity, the value of saturation current is larger.

By further exploiting the novel concept introduced by Planck, Einstein guessed that *a monochromatic e.m. wave of frequency ν consists of light quanta* which are described as massless pseudo-particles, travelling at the speed of light, and each carrying an energy

$$E^{\text{photon}} = h\nu \tag{1.16}$$

where h is given by equation (1.15). Light quanta are commonly named *photons*. According to the Einstein picture, *an e.m. wave is nothing other than a stream of photons*: its intensity is a direct measure of the number of light quanta there present. This hypothesis is just enough to explain the full set of experimental findings: (i) electrons are photoemitted (and, therefore, give rise to a current) whenever a photon with suitable energy is absorbed and transfers its energy to an electron of the metal cathode; (ii) photoemission can only occur if the photon energy is larger than the metal work function[15] W, i.e., if $\nu > W/h$ which for the typical values of W indeed corresponds to ultraviolet frequency; (iii) by increasing the light intensity we get a larger saturation current simply because a larger number of photons is absorbed and, consequently, electrons are emitted at a higher rate. Finally, the increasing/decreasing current behaviour under positive/negative bias is accounted for by the different bending of the photoelectron trajectories.

We conclude this digression by stressing the major conceptual step forward here outlined: *compelling experiments exist, proving that quantization is at work when considering interaction between matter and radiation*. This non-classical notion will be fully exploited in the next section.

1.2.3 Back to atomic spectra: the Bohr model

The *multiple* and *discrete* nature of atomic emission spectra has been already invoked twice to invalidate, respectively, the Thomson model and the classical

---

[14] While, despite its intensity, a radiation at any lower frequency is unable to generate a current.

[15] The work function of a metal (or, in general, of a solid) is the minimum thermodynamic work needed to remove an electron from it.

treatment of the Rutherford model. It is worth discussing in more detail the main spectral experimental findings to the aim of getting around the problems we have bumped into.

The first experimental evidence that atoms can absorb or emit e.m. energy was reported in 1859 by G Kirchhoff and R Bunsen. It is possible to measure *emission spectra* e.g., by recording the light emitted by a heated gas sample made of the same atoms and the *absorption spectra* by collimating a continuous light beam on a similar gas and detecting the transmitted signal. These experiments prove that: (i) the absorption/emission spectrum of any atom is specific of that atom, thus allowing unambiguous identification of any chemical species; (ii) both absorption and emission spectra consist in series of discrete lines; (iii) absorption and emission lines of any given atom fall at the very same wavelengths.

In the specific case of hydrogen, spectral lines are found at wavelengths supplied by the phenomenological *Rydberg equation*

$$\frac{1}{\lambda} = \mathcal{R}_H \left(\frac{1}{n_1^2} - \frac{1}{n_2^2} \right) \tag{1.17}$$

where $\mathcal{R}_H = 109\,678$ cm^{-1} is known as *Rydberg constant*. The n_1 and n_2 integer numbers identify the *spectral series* according to the following combination

$$\begin{aligned}
& n_1 = 1 \text{ and } n_2 = 2, 3, 4,\dots \text{ Lyman series} \\
& n_1 = 2 \text{ and } n_2 = 3, 4, 5,\dots \text{ Balmer series} \\
& n_1 = 3 \text{ and } n_2 = 4, 5, 6,\dots \text{ Paschen series} \\
& n_1 = 4 \text{ and } n_2 = 5, 6, 7,\dots \text{ Brackett series} \\
& n_1 = 5 \text{ and } n_2 = 6, 7, 8,\dots \text{ Pfund series}
\end{aligned} \tag{1.18}$$

where each series is named after the scientist that first discovered it. The equation (1.17) holds for both absorbed or emitted light, consistently with experiments.

In order to allow the nuclear H atom to honour equation (1.17) we need to add some new hypotheses going beyond the unsuccessful classical model. They have been elaborated in 1913 by N Bohr who further exploited the new quantum approach pioneered by Planck and Einstein. The first step was to postulate that the electron could move only along those planar and circular orbits for which *its angular momentum is an integer multiple of* $\hbar = h/2\pi$. According to the first Bohr postulate, the modulus L of the electron angular momentum is therefore given by[16]

$$L = m_e v r_n = n\hbar \tag{1.19}$$

where r_n is the radius of the orbit labelled by the integer number $n = 1, 2, 3,\dots$ and v is the electron speed. The orbits allowed by equation (1.19) are named *stationary states* since Bohr further assumed that *no emission of e.m. waves occurs when the electron is in such stable orbits.*

---

[16] We remark that the direction and orientation of the angular momentum are set by the planar geometry and by the direction of the electron motion.

For a stationary orbit of radius r we can identify the centripetal force with the Coulomb one

$$\frac{m_e v^2}{r} = \frac{1}{4\pi\epsilon_0}\frac{e^2}{r^2} \qquad (1.20)$$

and by combining this identity with equation (1.19) we get

$$r_n = \frac{h^2\epsilon_0}{\pi m_e e^2}\, n^2 = a_0\, n^2 \qquad (1.21)$$

where we have defined the *Bohr radius*

$$a_0 = \frac{h^2\epsilon_0}{\pi m_e e^2} = 0.529 \text{ Å} \qquad (1.22)$$

already used in section 1.2.1. Equation (1.21) represents a truly remarkable and intriguing result, since it is at odds with classical physics predictions: *the radius of the allowed stable orbits can only assume discrete or quantized values.* This is highlighted by the subscript appearing in the symbol r_n and n is referred to as the *quantum number*. The orbit corresponding to $n = 1$ is named *ground state*, while for any $n > 1$ we refer to *excited states*. In the limit $n \to +\infty$ the hydrogen atom no longer exists as a bound electron–proton system.

Let us now calculate the electron energy E_n for the nth stationary state. We have

$$E_n = \frac{1}{2}m_e v^2 - \frac{1}{4\pi\epsilon_0}\frac{e^2}{r_n} \qquad (1.23)$$

where, for obvious reasons, the same subscript used for the radius has been added to the energy as well. By calculating the electron velocity from equation (1.19) we easily get

$$E_n = -\frac{m_e e^4}{8\epsilon_0^2 h^2}\frac{1}{n^2} = -13.6 \text{ eV} \frac{1}{n^2} \qquad (1.24)$$

which states that *the energies of the stationary states are discrete or quantized.* In particular, for the ground state $n = 1$, equation (1.24) accurately provides the ionization energy of the hydrogen atom as known experimentally. The quantity

$$\frac{m_e e^4}{8\epsilon_0^2 h^2} = 13.6 \text{ eV} = 1 \text{ Ry} \qquad (1.25)$$

is referred to as the *Rydberg unit of energy* (or, simply, Rydberg) with symbol Ry.

A second postulate is needed to account for the atomic spectra, namely: *e.m. radiation is absorbed or emitted only when the electron undergoes a transition between stationary states.* More specifically: in a transition occurring between an initial stationary state described by the number n_{init} and a final one described by n_{fin}, we have absorption if $n_{\text{fin}} > n_{\text{init}}$ while we have emission if $n_{\text{fin}} < n_{\text{init}}$. They respectively correspond to the *absorption or emission of a photon* with energy given by the difference $E_{\text{photon}} = |E_{\text{init}} - E_{\text{fin}}|$. The frequency ν of the absorbed or emitted radiation for transitions occurring between the same states is calculated as

$$\nu = \frac{|E_{\text{init}} - E_{\text{fin}}|}{h} \tag{1.26}$$

which nicely explains the fact that absorption and emission lines fall at the same frequency. The observed discreteness of the atomic spectra is easily explained by calculating explicitly the wavelength of the absorbed/emitted photons

$$\frac{1}{\lambda} = \frac{\nu}{c} = \frac{|E_{\text{init}} - E_{\text{fin}}|}{ch} = \frac{m_e e^4}{8\epsilon_0^2 h^3 c} \left| \frac{1}{n_{\text{init}}^2} - \frac{1}{n_{\text{fin}}^2} \right| \tag{1.27}$$

which also provides a theoretical estimation of the hydrogen Rydberg constant

$$\mathcal{R}_{\text{H}} = \frac{m_e e^4}{8\epsilon_0^2 h^3 c} = 109\,737 \text{ cm}^{-1} \tag{1.28}$$

in rather good agreement with the experimental value. The resulting Bohr picture for the allowed transitions (both in emission and in absorption) involving the first four stationary states is illustrated in figure 1.1, where the same labelling as in equation (1.18) is used. In fact, the agreement could even be improved by observing that the Bohr model has implicitly assumed an infinite nuclear mass. Actually, the proton mass is 'only' 1836 times larger than the electron mass: this implies that both particles rotate around the centre-of-mass of the bound system. The use of the reduced mass value for the electron–proton system in equation (1.28) is therefore requested. By a simple calculation we get the corrected value 109 681 cm^{-1}, almost identical to the experimental one[17].

In conclusion, *the atomistic picture imposed by experimental evidence does require using the non-classical idea of quantization in order to explain other laboratory findings.* Indeed a rather engaging situation.

1.3 Subtleties of the quantum picture

1.3.1 The correspondence principle

Although the Bohr model succeeds in explaining the discrete nature of the hydrogen spectrum, it remains to be clarified what relationship exists between the *quantum frequencies* it predicts by combining equation (1.17) with equation (1.28)

$$\nu_{\text{Bohr}} = c\mathcal{R}_{\text{H}} \left(\frac{1}{n_1^2} - \frac{1}{n_2^2} \right) \tag{1.29}$$

and the *classical rotational frequencies* of the electron on the stationary orbits

---

[17] The same approach, namely the use of a reduced mass taking into account the finite value of the nuclear mass, can be followed in order to explain the experimentally observed *isotope shift* which accounts for the small, but measurable, difference in the spectra of hydrogen and deuterium.

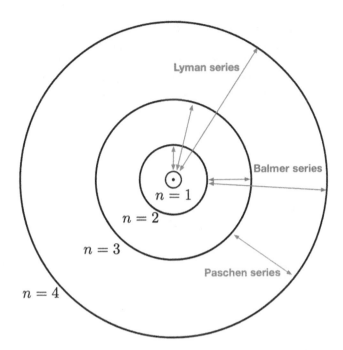

Figure 1.1. The first four $n = 1, 2, 3, 4$ Bohr orbits of the hydrogen atoms are shown, together with the corresponding emission/absorption lines for the Lyman (red), Balmer (blue), and Paschen (magenta) transitions.

$$\nu_{\text{classical}} = \frac{v}{2\pi r_n} = 2c\mathcal{R}_{\text{H}}\frac{1}{n^3} \tag{1.30}$$

that we obtain by using equation (1.19) for the linear velocity of rotation and equation (1.21) for the corresponding orbital radius r_n.

In searching for such a relationship, let us consider the specific case where $n_{1,2} \gg 1$ and $\Delta n = n_2 - n_1 \ll n_{1,2}$. Then, equation (1.29) can be manipulated as follows

$$\nu_{\text{Bohr}} = c\mathcal{R}_{\text{H}}\left(\frac{1}{n_1^2} - \frac{1}{n_2^2}\right) = c\mathcal{R}_{\text{H}}\frac{(n_2 - n_1)(n_2 + n_1)}{n_1^2 n_2^2} \simeq 2c\mathcal{R}_{\text{H}}\frac{\Delta n}{n_2^3} \tag{1.31}$$

which, by comparison to equation (1.30), leads to

$$\nu_{\text{Bohr}} = \Delta n\, \nu_{\text{classical}} \tag{1.32}$$

indeed a very intriguing result basically stating that for large enough quantum numbers (a condition which also allows to set $\Delta n = 1$), the emission/absorption spectral frequencies predicted by the Bohr model coincide with the classical frequencies of rotation.

This conclusion is generalized into the renowned *correspondence principle*, firstly enunciated by Bohr in 1923: *the quantum theory predictions must recover the predictions of classical physics whenever the quantum numbers describing the state of the system are very large.* It is instructive to remark, however, that this principle does not apply to hydrogen states characterized by a small value of the quantum number n; this implies that the present formulation of the quantum theory can be forced to agree with experiments and (in some respects) to be consistent with classical physics, but it is inherently patched up.

1.3.2 A common framework for the quantization rules

In equations (1.13) and (1.19) we have independently introduced two specific quantization rules, following the original ideas respectively formulated by Planck and Bohr. While a justification for their use can be heuristically found in the success obtained when addressing the blackbody radiation and the spectrum of atomic hydrogen, their independent formulation fails to provide a common and robust foundation which, in fact, was only formulated in 1916 as a general quantization condition by W Wilson and A Sommerfeld.

Following the formalism of classical mechanics, let us consider a general coordinate q with its associated general momentum p_q. If we further assume that *q is periodic in time*, then the *Wilson–Sommerfeld quantization condition* is cast in the form

$$\oint p_q \, dq = nh \tag{1.33}$$

where n is a suitable *quantum number* to be determined case by case and it is understood that the integration is taken over one period of variation of the q coordinate[18].

Let us at first apply the Wilson–Sommerfeld condition to the case of a one-dimensional harmonic oscillator for which any instantaneous state of motion is represented by a point in the two-dimensional phase-space having coordinates q, corresponding to the position, and p_q, corresponding to the linear momentum [11]. The trajectory followed by such a system in this space is an ellipse whose semiaxes a and b are

$$a = \sqrt{2E/\gamma} \quad b = \sqrt{2mE} \tag{1.34}$$

where

$$E = \frac{p_q^2}{2m} + \frac{1}{2}\gamma q^2 \tag{1.35}$$

is the total energy of the oscillator, m its mass and γ the force-constant of its restoring force. During a single period of oscillation, the representative point of the system travels just once around the entire ellipse. This implies that the integral appearing on

---

[18] By recalling some basic definition of classical mechanics [11], equation (1.33) is also referred to as the *quantization of the action.*

the left hand side of equation (1.33) is easily calculated as the area πab of the ellipse itself

$$\oint p_q \, dq = \frac{2\pi E}{\sqrt{\gamma/m}} = \frac{E}{\nu} \tag{1.36}$$

where we have introduced the oscillator frequency $2\pi\nu = \sqrt{\gamma/m}$. By exploiting the quantization condition reported on the right hand side of equation (1.33) we immediately get the Planck quantization of the energy of the harmonic oscillator provided in equation (1.13).

Let us now move to consider the more complex case of the hydrogen atom, which we will address at first as if it was a classical system formed by a fixed proton around which an electron moves along planar *elliptic trajectories*[19]. In order to describe the two-dimensional electron motion it is convenient to adopt the set polar coordinates $\{r, \theta\}$ shown in figure 1.2 (left). Then, the quantization condition given in equation (1.33) takes the twofold form

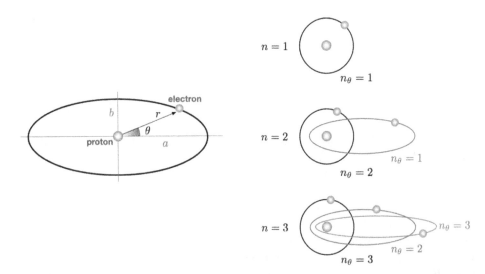

Figure 1.2. Left: the set of polar coordinates $\{r, \theta\}$ used in the Sommerfeld theory for the hydrogen atom; a and b are the major and minor semiaxis of the elliptic orbit, respectively. Right: pictorial representation of the electron orbits predicted by the Sommerfeld theory for the first three quantum states of the hydrogen atom. For the excited states ($n = 2$ and $n = 3$) the degenerate (in energy) elliptic orbits are shown in different colors, differing by the azimuthal quantum number n_θ. Figure is not in scale.

---

[19] The trajectories are planar simply because the only electron–proton interaction at work in this problem is provided by the Coulomb force, which is central. They are guessed elliptic just by a straightforward generalization of the Bohr model introduced by Sommerfeld.

$$\oint L \, d\theta = n_\theta h \qquad \oint p_r \, dr = n_r h \qquad (1.37)$$

where L and p_r are the angular and linear momentum associated with the θ and r variable, respectively. Since L is constant, the first condition leads to

$$2\pi L = n_\theta h \qquad (1.38)$$

which actually corresponds to the first Bohr postulate provided in equation (1.19). On the other hand, the second condition leads to[20]

$$L\left(\frac{a}{b} - 1\right) = n_r h \qquad (1.39)$$

where a and b are the major and minor semiaxis of the ellipse, as shown in figure 1.2 (left). For the trajectory to be stable, it must hold the mechanical stability condition given in equation (1.20). In conclusion, we have two quantization conditions and one stability condition providing three equations for the two geometrical variables a and b and for the energy E of the hydrogen atom. By solving this set of equations we easily obtain

$$a = \frac{h^2 \epsilon_0}{\pi m_e e^2} n^2 = a_0 \, n^2 \qquad b = \frac{n_\theta}{n} a \qquad E_n = -\frac{m_e e^4}{8 \epsilon_0^2 h^2} \frac{1}{n^2} = -\frac{1}{n^2} \mathrm{Ry} \qquad (1.40)$$

where we have introduced the quantum number $n = n_\theta + n_r$. Since $n_\theta = 1, 2, 3, \ldots$ and $n_r = 0, 1, 2, 3, \ldots$, then we have $n = 1, 2, 3, \ldots$; furthermore, for any given value of n, it must hold $n_\theta = 1, 2, 3, \ldots , n$. Usually n and n_θ are referred to as the *principal* and *azimuthal* quantum number, respectively.

The Sommerfeld theory for the hydrogen atom has obtained two important results. First of all, it reproduces the quantization of the orbits and of the energy in a more general way than in the Bohr model, following a unified quantization condition which also leads to the discretization of the energy of a harmonic oscillator; it also predicts the existence of both circular and elliptic obits. Next, it introduces the unprecedented (and rather subtle) quantum concept of *degeneracy*: while the energy of a state only depends on the principal quantum number, the shape of the corresponding orbit actually depends on the azimuthal quantum number which, for any assigned n, can assume several different values. In other words, the electron could accomodate on differently-shaped trajectories, still keeping the same energy. The situation is shown in figure 1.2 (right) for the three quantum states. The existence of degenerate states has a very important consequence in atomic spectroscopy: *it must be admitted that a single spectroscopic line could correspond to several possible transitions between pairs of states with same energy difference, but differing by other physical features.* This very intriguing physics will be extensively discussed in chapters 3 and 4.

---

[20] It is worth remarking that this result justifies the Sommerfeld assumption of elliptic trajectory: for a circular one the two semiaxes are just the same and, therefore, the quantization rule here given is not applicable.

1.3.3 More on degeneracy: space quantization

In the previous section, we treated the motion of the electron in terms of just two degrees of freedom, describing its position in the plane of motion. We can now complete our physical picture by introducing a third variable describing the orientation of this plane with respect to a generic direction in space. This variable corresponds to the angle ϕ formed between the normal to the plane and the chosen direction and, similarly to the two variables already introduced, is periodic. Therefore, three Wilson–Sommerfeld quantization conditions are now defined

$$\oint p_r \, dr = n_r h \qquad \oint p_\theta \, d\theta = n_\theta h \qquad \oint p_\phi \, d\phi = n_\phi h \qquad (1.41)$$

where p_r, p_θ, and p_ϕ are the three conjugated momenta.

By writing the energy of the electron in terms of the above three variables and by setting it equal to the energy previously calculated just in terms of the r and θ variables[21], it is not difficult to show that the new quantum number n_ϕ is restricted to values $n_\phi = \pm 1, \pm, 2, \pm 3, \ldots, \pm n_\theta$. This condition imposes the so-called *space quantization* rule to the electron orbit, stating that it can assume *only discrete orientations with respect to any arbitrarily chosen direction*.

Even in this case, the new quantum number does not enter in the formula for the energy, which remains as in equation (1.40). In other words, we do have a second kind of degeneracy which can only be removed by reducing the symmetry of the problem from spherical to axial. Physically this can be achieved by applying an external magnetic field to the hydrogen atom and, for this reason, n_ϕ was labelled as *magnetic quantum number*. By resolving the degeneracy in n_ϕ the emission/absorption spectra become much more rich (that is, they show much more spectral lines), a situation which was referred to as the *fine structure* of hydrogen spectrum: it will be thoroughly discussed in section 3.3.

1.4 The dual nature of physical phenomena

1.4.1 Wave–matter duality

It is a matter of common experience that macroscopic physical phenomena reveal *either* as waves *or* as particles. By contrast, the Einstein hypothesis about the nature of e.m. waves as a flux of photons challenges this (pre)conception, which we now understand to be only due to the limits of our sensorial experience of the physical word. Light can in fact manifest *either* as a wave *and* as a beam of (pseudo)particles, according to the actual phenomenon we are addressing. For sure the physics of our ocular vision or the propagation of light in vacuum are phenomena very well described by wave equations. On the other hand, the absorption/emission of light by an atomic system or the photoemission of electrons from a metal plate can only be explained by invoking the concept of photon.

---

[21] The total energy of the free hydrogen atom is of course not affected by introducing possible rotations of the planar orbit with respect to any arbitrary direction in space.

In principle, we could speculate that this duality is similarly valid for massive particles, as first discussed by L de Broglie in 1924. If for a photon we can relate wave-like and particle-like properties through such relations as $E = h\nu$ or $\mathbf{p} = \hbar\mathbf{k}$ (where $\mathbf{p}$ is the photon momentum and $\mathbf{k}$ is the wavevector of the corresponding e.m. wave), then we could guess that a *matter wave* of wavelength λ is associated to any particle with mass m and moving with velocity v according to

$$\lambda = \frac{h}{p} = \frac{h}{mv} = \frac{h}{\sqrt{2mE_{\text{kin}}}} \tag{1.42}$$

where $\mathbf{p} = mv$ is the particle momentum and $E_{\text{kin}} = mv^2/2$ is its kinetic energy. This statement is nothing other than speculation if no experimental evidence is supplied to support it.

Interestingly enough, we have a rather compelling laboratory proof [1, 2, 13]: a beam of electrons is diffracted by a crystalline sample giving rise to constructive/destructive interference phenomena as if they were in fact waves. This experience, first performed by C L Davisson and L H Germer in 1926, provided a direct confirmation that the wavelength of diffracted electrons was exactly that predicted by de Broglie. So, electrons can be safely described as massive particles when studying electric phenomena (electrostatics or even the photoelectric effect), but are better described as matter waves when diffracted by a crystal. Over the years, a large number of additional experiments proved that microscopic massive objects behave as waves (for instance, diffraction and interference have been observed in systems consisting of atom beams).

It is worth remarking that the corpuscular character of e.m. radiation and, similarly, the wave-like character of particles are best observed in microscopic phenomena, i.e., in those phenomena occurring at the atomic length scale (or bottom)[22]. On the other hand, the more familiar treatment of a radiation field as a wave and of a massive particle as a corpuscle is more appropriate on the human length scale. In an attempt to reconcile the ordinary physical description to the emerging new picture, in 1927 N Bohr formulated the *complementarity principle*, in that: *the wave-like and particle-like descriptions of the same phenomenon are complementary*. If an experiment states the corpuscular character of, say, radiation, then it is impossible to probe its wave character by the same measurement, and conversely. The same holds for particles. In an equivalent phrasing, we can state according to Bohr that *it is the specific measurement performed that makes either the corpuscular or the wave character emerge*.

The wave-like character of an atomic electron is easily reconciled with the Bohr model for hydrogen. If such an electron is described through its de Broglie wavelength λ, then any stationary state must be described by *a standing matter wave along the corresponding orbit* since the electron is bound to the nucleus. This implies that the circular radius r must obey the condition

---

[22] For instance, if we consider an electron accelerated by a 10^4 V electrostatic potential (a situation typical of cathode-ray tubes in old television sets), equation (1.42) gives a wavelength as small as $\sim 10^{-11}$ m: a characteristic which is really very hard to observe at the macroscopic scale.

$$2\pi r_n = n\lambda \tag{1.43}$$

with $n = 1, 2, 3,...$ labelling the stationary states. By inserting this value of r_n into equation (1.20) and by making use of the de Broglie relation cast in equation (1.42) we easily get equation (1.21). This result works as well as a heuristic justification of the quantization postulate on the electron angular momentum.

1.4.2 A constitutive equation for matter waves

Once we acknowledge that any microscopic particle, say an electron, behaves as a matter wave, we must duly feel committed in the search for the constitutive equation ruling over the physics of such unfamiliar waves. This is indeed a very subtle problem that, as a matter of fact, is still an open issue of intense fundamental research. We can nevertheless draw a semi-empirical picture which could be the conceptual guideline in developing a more satisfying formal theory.

As a first step we recognize that any wave, of whichever nature, is described by the d'Alembert equation. Accordingly, if we name $\Psi(x, t)$ *the wavefunction* of the matter wave describing an electron in one-dimensional motion with speed v, we can write

$$\frac{\partial^2\Psi(x, t)}{\partial x^2} - \frac{1}{v^2}\frac{\partial^2\Psi(x, t)}{\partial t^2} = 0 \tag{1.44}$$

where x indicates the direction of motion. By assuming an harmonic time dependence $\Psi(x, t) = \psi(x)\exp(i\omega t)$ we easily get

$$\frac{d^2\psi(x)}{dx^2} + \left(\frac{\omega}{v}\right)^2\psi(x) = 0 \tag{1.45}$$

where $\omega = vk = 2\pi v/\lambda$ and $k = 2\pi/\lambda$ is the wavenumber of the matter wave with de Broglie wavelength $\lambda = h/m_e v$. Equation (1.45) is easily rewritten in the more useful form

$$\frac{d^2\psi(x)}{dx^2} + \left(\frac{m_e v}{\hbar}\right)^2\psi(x) = 0 \tag{1.46}$$

which is really intriguing since *it contains a juxtaposition of wave-like and particle-like features*: a wavefunction indeed appears together with a mechanical linear momentum. Let us further exploit the corpuscular feature by assuming that the electron is subjected to an external potential $V(x)$, so that its total energy is $E = m_e v^2/2 + V(x)$. Eventually, equation (1.46) is cast in the form

$$\frac{d^2\psi(x)}{dx^2} + \frac{2m_e}{\hbar^2}[E - V(x)]\psi(x) = 0 \tag{1.47}$$

or even better

$$-\frac{\hbar^2}{2m_e}\frac{d^2\psi(x)}{dx^2} + V(x)\psi(x) = E\psi(x) \tag{1.48}$$

which represents *the constitutive equation for the matter wave* describing our electron.

By solving this equation, once we know the potential $V(x)$ and the boundary conditions for the specific problem we are interested in, we obtain quite a bit of information, namely: the energy E of the particle and the wavefunction $\psi(x)$ describing its physical state. The locution 'physical state' is admittedly a bit vague at this level. While it will be fully clarified when developing the formal quantum theory, here we can accept its meaning as: if we knew the wavefunction $\psi(x)$, then we could calculate any physical property of the electron represented by such a matter wave. The total amount of information so acquired defines the electron physical state.

An important open question is raised by equation (1.48): *which is, if any, the physical meaning of $\psi(x)$?* The answer is not that easy since, for any $V(x)$ actually describing a physical situation, it turns out that the $\psi(x)$ is a complex, single-valued function of the real variable x: it is not possible to directly attach any physical meaning to a complex quantity, since it cannot be addressed by a laboratory measure.

An effective way to provide a meaningful significance of the wavefunction consists in reconsidering the dual nature of e.m. waves. According to Einstein, the wave intensity is somehow a measure of the number of photons, as experimentally proved in the photoelectric effect by direct link existing between the saturation current and the intensity of the impinging ultraviolet radiation. On the other hand, the intensity of such a wave is given by its square amplitude. Since we accepted that both the wave-like and particle-like descriptions of an e.m. are true, we can conclude that *the square amplitude of the e.m. wave in any given point of the space is directly related to the probability of finding photons in that point.*

We extend by analogy such a conclusion and state that for any matter wave $\psi(\mathbf{r})$ describing a given particle, *the quantity $|\psi(\mathbf{r})|^2 \, d\mathbf{r}$ represents the probability of finding that particle in the infinitesimal volume $d\mathbf{r}$ centred at point $\mathbf{r}$*. This conclusion was fully formalized and generalized in 1926 by M Born who provided the foundation of the statistical interpretation of quantum mechanics, well verified by and fully consistent with all further theoretical developments.

Further reading and references

[1] Bransden B H and Joachain C J 1983 *Physics of Atoms and Molecules* (Harlow: Addison-Wesley Longman)

[2] Demtröder W 2010 *Atoms, Molecules and Photons* (Heidelberg: Springer)

[3] Glazer M and Wark J 2001 *Statistical Mechanics: A Survival Guide* (Oxford: Oxford University Press)

[4] Colombo L 2022 *Statistical Physics of Condensed Matter Systems Physics: A Primer* (Bristol: IOP Publishing)

[5] Swendsen R H 2012 *An Introduction to Statistical Mechanics and Thermodynamics* (Oxford: Oxford University Press)

[6] Einstein A 1956 *Investigations on the Theory of Brownian Motion* (New York: Dover)

[7] Egerton R F 2010 *Physical Principles of Electron Microscopy* (New York: Springer Science+Business Media)

[8] Haugstad G 2010 *Atomic Force Microscopy* (Hoboken, NJ: Wiley)

[9] Feynman R P, Leighton R B and Sands M 1963 *The Feynman Lectures on Physics* (Reading, MA: Addison-Wesley)

[10] Atkins P and De Paula J 2010 *Physical Chemistry* 9th edn (Oxford: Oxford University Press)

[11] Goldstein H 1996 *Classical Mechanics* 2nd edn (Reading, MA: Addison-Wesley)

[12] Povh B, Rith K, Scholz C and Zetsche F 2009 *Particles and Nuclei* 6th edn (Heidelberg: Springer)

[13] Eisberg R and Resnick R 1985 *Quantum Physics of Atoms, Molecules, Solids, Nuclei, and Particles* 2nd edn (Hoboken, NJ: Wiley)

[14] Miller D A B 2008 *Quantum Mechanics for Scientists and Engineers* (New York: Cambridge University Press)

[15] Griffiths D J and Schroeter D F 2018 *Introduction to Quantum Mechanics* 3rd edn (Cambridge: Cambridge University Press)

[16] Bransden B H and Joachain C J 2000 *Quantum Mechanics* (Englewood Cliffs, NJ: Prentice-Hall)

IOP Publishing

Atomic and Molecular Physics (Second Edition)
A primer
Luciano Colombo

Chapter 2

Essential quantum mechanics

Syllabus—The rudiments of the non-relativistic quantum theory are here worked out, in preparation for its thorough application to atomic and molecular physics developed in the next chapters. While the formal architecture will rely on a number of statements, definitions, and mathematical properties introduced either for self-consistency and for further convenience, its physical foundation is found for comparison to the phenomenological and conceptual achievements discussed in the previous chapter. All properties here reported—although representing just a summary of quantum mechanics—are discussed together with their formal proof, so as to provide good training for getting familiar with the quantum formalism.

2.1 The wavefunction

2.1.1 Definition

The *state* of a non-relativistic particle system is defined by the entirety of its *measurable physical properties*. In quantum mechanics a *wavefunction* Ψ, also referred to as *state vector*, is associated with any such physical state. Its definition is manifold:

- Ψ is a complex, single-valued, and differentiable (and therefore continuous) function of all the particle positions $\mathbf{r}_i$ (where i labels the particles) and the time t;
- no quantum state is associated with the null wavefunction or, equivalently, if the system cannot occupy a given state, then the corresponding wavefunction is $\Psi = 0$;
- Ψ is calculated by solving the *Schrödinger equation*, namely the postulated constitutive quantum mechanical equation;
- through a system of formal rules to be defined, the knowledge of Ψ will allow one to predict the actual value of any property when the system occupies that state; such predicted values can be directly compared to the experimental measurements of the corresponding physical quantities.

doi:10.1088/978-0-7503-5734-0ch2

We remark that the link to the early physics of quanta outlined in chapter 1 is the *identification of wavefunctions as the formal counterpart of matter waves*. While, according to de Broglie's treatment, the latter ones describe particles, here we rather state that Ψ *describes the physical state of a particle*.

2.1.2 Properties

If at time t a physical system can possibly occupy two different states, respectively described by the wavefunctions Ψ_1 and Ψ_2, then at the same time *it can also occupy any state described by the combined wavefunction*

$$\Psi = a_1\Psi_1 + a_2\Psi_2 \tag{2.1}$$

with $a_{1,2}$ arbitrary complex numbers[1]. This assumption is called *superposition principle*, which also implies that Ψ and $a\Psi$ ($a \neq 0$ is any arbitrary complex number) both describe the very same state of the system. The superposition principle dictates that *the Schrödinger equation must be linear*.

By considering a single-particle system, it is assumed that the quantity[2]

$$|\Psi|^2 \, d\mathbf{r} = \Psi^*\Psi \, d\mathbf{r} \tag{2.2}$$

is proportional to *the probability of finding at time t the particle in the infinitesimal volume $d\mathbf{r}$ centred at position* $\mathbf{r}$. Furthermore, if for a given state the function $|\Psi'|^2$ is integrable with

$$\int_\Omega |\Psi'|^2 \, d\mathbf{r} = A \tag{2.3}$$

where A is a real number and Ω represents *the whole space accessible to the particle*, then we will more conveniently write the corresponding state vector as $\Psi = A^{-1/2}\Psi'$ so that

$$\int_\Omega |\Psi|^2 \, d\mathbf{r} = 1 \tag{2.4}$$

which is known as the *normalization condition*. In this way $|\Psi|^2$ calculated in position $\mathbf{r}$ at time t provides the *probability density* of finding the particle in that point at that time. This *probabilistic interpretation of the quantum wavefunction* is clearly inspired by the previous achievements of the early physics of quanta (see chapter 1) and, if

---

[1] However, it must be observed that for a system consisting of *two or more identical particles*, not all the linear combinations of arbitrary solutions of the Schrödinger equation are allowed, since some symmetry constraints must be obeyed. This situation of great relevance for multi-electron atoms will be explicitly addressed later in this chapter.

[2] The asterisk * will always indicate the complex conjugate function.

added to the continuity of Ψ, implies that the probability density is always continuous[3].

Unless otherwise stated, *we will always assume normalized wavefunctions*. This mathematically implies that any state vector is defined up to a phase factor $e^{i\alpha}$ where α is an arbitrary real number; it will be proved in the following that this will not drive to any ambiguity in the physical description of the associated quantum system.

2.2 Quantum operators

2.2.1 Definition

We agree upon setting a *quantum operator* $\hat{F}$ for any physical quantity F in such a way that its *expectation value*

$$\langle f \rangle = \int \Psi^* \hat{F} \Psi d\mathbf{r} \tag{2.5}$$

represents *the mean value of such a quantity* when the system is an arbitrary state described by the wavefunction Ψ. Such a mean value has an experimental counterpart: it can be understood as the value obtained by averaging over many different experimental measurements.

While in general we guess that the quantum operator $\hat{F}$ requires either algebraic or differential operations on Ψ, we still miss both a procedure to build it and the full control of its properties. This the topic of the present section.

2.2.2 Building a quantum operator

In order to fulfil the superposition principle given in equation (2.1), *a quantum operator must be linear*

$$\hat{F}(a\Psi) = a\hat{F}\Psi \text{ and } \hat{F}(\Psi_1 + \Psi_2) = \hat{F}\Psi_1 + \hat{F}\Psi_2 \tag{2.6}$$

where a is any complex number. Furthermore, if F is a physical quantity *its observable value* must be real or, equivalently, it must hold that

$$\langle f \rangle = \langle f \rangle^* \rightarrow \int \Psi^* \hat{F} \Psi d\mathbf{r} = \int \Psi \hat{F}^* \Psi^* d\mathbf{r} \tag{2.7}$$

a condition stating that $\hat{F}$ is an *Hermitian operator*.

Linear and Hermitian quantum operators are built by a two-step procedure, namely: the classical expression for the quantity F is at first written; next, the position vector $\mathbf{r}$ is replaced by the *multiplication by* $\mathbf{r}$, while the linear momentum vector $\mathbf{p}$ is replaced by the *differential operator* $-i\hbar\nabla$. It is easy to work out the correspondences between classical quantities and quantum operators, as shown in

---

[3] We remark that, in case Ψ describes an N-particle system and, therefore, depends on the spatial coordinates $\mathbf{r}_1, \mathbf{r}_2, \ldots, \mathbf{r}_N$ of all its constituents, then equation (2.2) must be generalized in that: $|\Psi|^2 d\mathbf{r}_1 d\mathbf{r}_2 \cdots d\mathbf{r}_N$ provides the probability that a direct measure finds the first particle in the volume $d\mathbf{r}_1$ centred at position $\mathbf{r}_1$, the second one in the volume $d\mathbf{r}_2$ centred at position $\mathbf{r}_2$, and so on.

Table 2.1. Some simple quantum operators of extensive use in this book. It must be understood that: $\vec{\nabla} = \frac{\partial}{\partial x}\vec{i} + \frac{\partial}{\partial y}\vec{j} + \frac{\partial}{\partial z}\vec{k}$, where $\{\vec{i}, \vec{j}, \vec{k}\}$ are three unit vectors identifying the Cartesian axes, and $\nabla^2 = \frac{\partial^2}{\partial x^2} + \frac{\partial^2}{\partial y^2} + \frac{\partial^2}{\partial z^2}$.

Physical quantity	Classical expression	Quantum operator
Position	$\mathbf{r}$	$\hat{\mathbf{r}} = \mathbf{r}$
Linear momentum	$\mathbf{p} = m\mathbf{v}$	$\hat{\mathbf{p}} = -i\hbar\vec{\nabla}$
Angular momentum	$\mathbf{L} = \mathbf{r} \times \mathbf{p}$	$\hat{\mathbf{L}} = -i\hbar\mathbf{r} \times \vec{\nabla}$
Kinetic energy	$T = \frac{p^2}{2m}$	$\hat{T} = -\frac{\hbar^2}{2m}\nabla^2$
Coulomb potential energy	$V = \frac{1}{4\pi\varepsilon_0}\frac{e^2}{r}$	$\hat{V} = \frac{1}{4\pi\varepsilon_0}\frac{e^2}{r}$

table 2.1, in a few relevant cases. For instance, the operator associated with the total energy of the hydrogen atom in the Bohr model is

$$\text{classic: } E = \frac{p^2}{2m} - \frac{1}{4\pi\varepsilon_0}\frac{e^2}{r} \rightarrow \text{quantum: } \hat{H} = -\frac{\hbar^2}{2m}\nabla^2 - \frac{1}{4\pi\varepsilon_0}\frac{e^2}{r} \quad (2.8)$$

where we have introduced the symbol $\hat{H}$ for the *Hamiltonian operator* hereafter used to label the quantum operator for the total energy of a physical system.

The product between two operators $\hat{F}$ ed $\hat{M}$ is defined as

$$(\hat{F}\hat{M})\Psi = \hat{F}(\hat{M}\Psi) \quad (2.9)$$

or, equivalently, by first applying the operator $\hat{M}$ to the state vector Ψ and next by applying the operator $\hat{F}$ to the resulting function. It is therefore plain that *the order of application is important* or, in other words, the commutation of two operators is not irrelevant. An additional tool is accordingly defined and named *commutator*

$$[\hat{F}, \hat{M}] = \hat{F}\hat{M} - \hat{M}\hat{F} \quad (2.10)$$

so that in general $[\hat{F}, \hat{M}]\Psi \neq 0$. On the other hand, when it holds that $[\hat{F}, \hat{M}]\Psi = 0$ then $\hat{F}$ and $\hat{M}$ are said to be *commuting operators*. In any case, the commutator of two Hermitian operators is an anti-Hermitian operator. By using table 2.1 it is an easy task to prove that

$$[\hat{r}_i, \hat{r}_j] = 0 \quad [\hat{p}_i, \hat{p}_j] = 0 \quad [\hat{r}_i, \hat{p}_j] = i\hbar\delta_{ij} \quad [\hat{L}_i, \hat{L}_j] = i\hbar\hat{L}_k \quad (2.11)$$

where i, j, k label the Cartesian components of the position, linear momentum, and angular momentum operators while $\delta_{ij} = 1$ only if $i = j$, otherwise it is zero.

2.2.3 Eigenfunctions and eigenvectors

Let us consider a physical quantity F and suppose that a direct measure provides the exact value f. Since quantum theory provides through equation (2.5) the corresponding average value $\langle f \rangle$ ideally taken over many other independent evaluations, we can define the *mean square deviation* $\langle (\Delta f)^2 \rangle$

$$\langle (\Delta f)^2 \rangle = \int \Psi^*(f - \langle f \rangle)^2 \Psi \, d\mathbf{r} = \int |(f - \langle f \rangle)\Psi|^2 \, d\mathbf{r} \tag{2.12}$$

of the quantity F when the system is a state described by the wavefunction Ψ. In quantum physics it is of paramount importance to focus on those states for which it is found $\langle (\Delta f)^2 \rangle = 0$, a very peculiar situation where *the expectation value exactly matches the measured one*. According to equation (2.12), in such states it holds

$$\hat{F}\Psi = f\Psi \tag{2.13}$$

which is commonly named *eigenvalue equation for the operator* $\hat{F}$. When equation (2.13) is fulfilled, it is said that Ψ *is an eigenstate of this operator* or, equivalently, that Ψ *is an eigenfunction of* $\hat{F}$, while f is the corresponding *eigenvalue* providing the *unique value* of the observable F for that state. In different words: when a system is in a physical state corresponding to an eigenstate of a given observable, then *any* measurement of this quantity will provide invariably the same value.

The set of solutions of the linear, homogenous eigenvalue equation for $\hat{F}$ is called *the spectrum of this operator*. If the different eigenvalues form a discrete set, than the spectrum is also said to be discrete. By referring to an *Hermitian operator with a discrete spectrum of real eigenvalues*, we cast the eigenvalue equation in the form

$$\hat{F}\Psi_n = f_n\Psi_n \tag{2.14}$$

where n labels both the eigenstates/-values and it is known as *quantum number*. In summary, until a system is in the state Ψ_n any experimental measurement of the quantity F will always provide the value f_n with no uncertainty, i.e., with zero mean square deviation.

In general it could happen that two (or more) linearly independent eigenfunctions of a given operator have the very same eigenvalue (like, e.g., the Hamiltonian operator of the hydrogen atom, which we repeatedly refer to since, so far, it is basically the only quantum system we are familiar with). This, in turn, could occur for more than one eigenvalue. We describe this situation by saying that the *spectrum of the operator is degenerate* and name *degree of degeneracy* the number of eigenfunctions associated with a given eigenvalue. On the other hand, if a one-to-one correspondence is found we describe the spectrum as non-degenerate.

The eigenfunctions of a given operator have some remarkable mathematical properties with important physical consequences. Since they will be extensively addressed when discussing atomic and molecular physics problems, we are outlining below some of them. While a thorough treatment can be found elsewhere [1–3], here we will only consider the specific (but common and relevant) case of *an Hermitian operator* $\hat{F}$ *with a discrete and non-degenerate spectrum* by adopting the notation introduced in equation (3.97) above.

2.2.3.1 Orthonormality

If $n \neq m$ then

$$\hat{F}\Psi_n = f_n\Psi_n \rightarrow \Psi_m^*\hat{F}\Psi_n = f_n\Psi_m^*\Psi_n$$

$$\hat{F}^*\Psi_m^* = f_m\Psi_m^* \rightarrow \Psi_n\hat{F}^*\psi_m^* = f_m\Psi_n\Psi_m^*$$

so that, thanks to the Hermitian character of $\hat{F}$, we easily obtain

$$0 = (f_n - f_m)\int\Psi_m^*\Psi_n d\mathbf{r} \tag{2.15}$$

which is satisfied only provided that

$$\int\Psi_m^*\Psi_n\, d\mathbf{r} = \delta_{nm} \tag{2.16}$$

a result proving that *the eigenfunctions of the operator $\hat{F}$ are orthonormal.*

2.2.3.2 Completeness

Equation (2.16) implies that the only wavefunction normal to any other one is $\Psi = 0$ and, accordingly, *the full assembly $\{\Psi_n\}$ forms a complete set of functions*: any other function Φ depending on the same variables, defined in the same portion of space, and there obeying the same normalization condition can be written as a linear combination

$$\Phi = \sum_n a_n\Psi_n \tag{2.17}$$

where the coefficients a_n are given by

$$a_n = \int\Psi_n^*\Phi\, d\mathbf{r} \tag{2.18}$$

because of orthonormality. If the state described by Φ is *not* an eigenstate of $\hat{F}$, then repeated measurements of the quantity F will provide different results with a mean value given by

$$\langle f\rangle = \sum_n f_n|a_n|^2 \tag{2.19}$$

where, thanks to normalization, it holds that

$$\int\Phi^*\Phi\, d\mathbf{r} = \sum_n|a_n|^2 = 1 \tag{2.20}$$

which is known as *completeness condition*[4].

---

[4] The conclusions here drawn under the assumption that the operator has a discrete and non-degenerate spectrum, can in fact be generalized to the case where one or more eigenvalues are degenerate. A more mathematically refined argument indeed proves that the set of eigenfunctions forms in any case a complete set, while orthonormality can be afterwards imposed by construction.

2.2.3.3 A remark

The combination of the orthogonality and completeness of a set of eigenfunctions allows one to work out a functional representation of the δ-Dirac function[5]

$$\sum_n \Psi_n^*(\mathbf{r}')\Psi_n(\mathbf{r}) = \delta(\mathbf{r}' - \mathbf{r}) \tag{2.22}$$

which will become very useful in many practical quantum mechanical calculations.

2.2.3.4 Commuting properties

The picture so far elaborated makes it clear that *whether a wavefunction is simultaneously an eigenfunction of two different operators, then both the corresponding observables have well-defined values* in that state. This is not at all a trivial conclusion, deserving full explanation.

Assuming a discrete non-degenerate spectrum for both $\hat{F}$ and $\hat{G}$ operators, let Ψ_n be a wavefunction such that

$$\hat{F}\Psi_n = f_n\Psi_n \;\rightarrow\; \hat{G}\hat{F}\Psi_n = f_n\hat{G}\Psi_n = f_n g_n\Psi_n$$

$$\hat{G}\Psi_n = g_n\Psi_n \;\rightarrow\; \hat{F}\hat{G}\Psi_n = g_n\hat{F}\Psi_n = g_n f_n\Psi_n$$

which leads to

$$(\hat{G}\hat{F} - \hat{F}\hat{G})\Psi_n = (f_n g_n - g_n f_n)\Psi_n = 0 \tag{2.23}$$

or, equivalently: $[\hat{G}, \hat{F}] = 0$. Therefore: *if two physical quantities have well-defined values for a system in a given state, then the corresponding operators commute.*

The opposite is true as well. Let us suppose that

$$[\hat{G}, \hat{F}] = 0 \;\; \text{with} \;\; \hat{G}\Psi_n = g_n\Psi_n \tag{2.24}$$

which leads to

$$\hat{F}(\hat{G}\Psi_n) = \hat{G}(\hat{F}\Psi_n) = g_n(\hat{F}\Psi_n) \tag{2.25}$$

proving that $\hat{F}\Psi_n$ is in fact an eigenfunction of $\hat{G}$ with eigenvalue g_n. Since the assumed features of the spectrum, $\hat{F}\Psi_n$ should differ from Ψ_n to within a suitable multiplicative factor that, for further convenience, we label f_n and formally it holds that

$$\hat{F}\Psi_n = f_n\Psi_n \tag{2.26}$$

---

[5] While the full mathematics of the δ-Dirac function is found in any advanced textbook of quantum mechanics, including those reported in the bibliography of the present chapter, we recall that its operational definition is

$$\int_\Omega \xi(\mathbf{r})\delta(\mathbf{r}' - \mathbf{r})d\mathbf{r} = \xi(\mathbf{r}') \tag{2.21}$$

where $\xi(\mathbf{r})$ is any continuous function defined over the space volume Ω containing $\mathbf{r}'$. This definition supports its description as a singular function vanishing for all points such that $\mathbf{r}' - \mathbf{r} \neq 0$.

In conclusion, *if two operators commute, then there exists a common set of eigenfunctions.*

2.3 The uncertainty principle

We have just proved in the previous section that two physical quantities cannot simultaneously have well-defined values in the same state if their corresponding operators do not commute. In this case it is, however, possible to establish an inequality ruling over the uncertainty affecting our knowledge of the values of the two observables.

Let us consider two non-commuting Hermitian operators $\hat{F}$ and $\hat{G}$ such that

$$[\hat{F}, \hat{G}] = i\hat{M} \tag{2.27}$$

where $\hat{M}$ is also Hermitian. This initial statement is inspired by equation (2.11). The corresponding expectation values on the state Ψ are calculated according to equation (2.5) so that we can introduce the following two operators

$$\hat{\Delta F} = \hat{F} - \langle f \rangle \quad \hat{\Delta G} = \hat{G} - \langle f \rangle \tag{2.28}$$

which obey the following commuting relation

$$[\hat{\Delta F}, \hat{\Delta G}] = i\hat{M} \tag{2.29}$$

as is easily verified by a simple calculation.

Let us now consider a real variable ξ and take into consideration the integral

$$\int | (\xi\hat{\Delta F} - i\hat{\Delta G})\Psi |^2 \, d\mathbf{r} \geqslant 0 \tag{2.30}$$

which can be straightforwardly transformed as

$$\int \psi^*(\xi\hat{\Delta F} + i\hat{\Delta G})(\xi\hat{\Delta F} - i\hat{\Delta G})\Psi d\mathbf{r} \geqslant 0 \tag{2.31}$$

since both $\hat{\Delta F}$ and $\hat{\Delta G}$ are as well Hermitian. The explicit calculation of the integral leads to

$$\langle (\Delta f)^2 \rangle \left[\xi + \frac{\langle m \rangle}{2\langle (\Delta f)^2 \rangle} \right]^2 + \langle (\Delta g)^2 \rangle - \frac{\langle m^2 \rangle}{4\langle (\Delta f)^2 \rangle} \geqslant 0 \tag{2.32}$$

with self-explaining symbols $\langle \Delta f \rangle$, $\langle \Delta g \rangle$, and $\langle m \rangle$: they simply are the eigenvalues of the corresponding operators defined in equations (2.27) and (2.28). In order to satisfy this inequality for all possible values of the arbitrary parameter ξ it must hold that

$$\langle (\Delta f)^2 \rangle \langle (\Delta g)^2 \rangle \geqslant \frac{\langle m^2 \rangle}{4} \tag{2.33}$$

which is known as the *Heisenberg relation* or *uncertainty principle*. It states a fundamental limit to the accuracy with which the values for certain pairs of physical quantities can be predicted.

In order to appreciate the physical consequences of the demonstrated uncertainty, let us consider the position and momentum non-commuting operators. By using equation (2.11) in combination with equation (2.33) we immediately get

$$\langle (\Delta r_i)^2 \rangle \langle (\Delta p_i)^2 \rangle \geqslant \frac{\hbar^2}{4} \tag{2.34}$$

which for the specific x-components of the position and momentum leads to

$$\Delta r_x \, \Delta p_x \geqslant \frac{\hbar}{2} \tag{2.35}$$

corresponding to the most celebrated form of the Heisenberg relation: for any given quantum state of our system, we cannot know simultaneously its position and momentum with absolute certainty.

This remarkable result can be extended to other two observables of key importance, like energy and time. We preliminarily observe that

$$[\hat{H}, t] = i\hbar \tag{2.36}$$

where, as explained in the next section, $\hat{H}$ is the quantum mechanical operator such that $\hat{H}\Psi = E\Psi$ where E is the total energy of a conservative system, while t is the time operator (actually corresponding just to the multiplication by t). Then, we easily get

$$\Delta E \Delta t \geqslant \frac{\hbar}{2} \tag{2.37}$$

corresponding to the second most famous form of the uncertainty principle: it relates the uncertainty ΔE in the determination of the system energy to the time lapse Δt needed to determine it.

2.4 Time evolution

2.4.1 The Schrödinger equation

It is postulated that the *constitutive quantum mechanical equation* obeyed by any wavefunction $\Psi(\mathbf{r}, t)$ describing the state of a given particle is

$$i\hbar \frac{\partial \Psi(\mathbf{r}, t)}{\partial t} = \hat{H}\Psi(\mathbf{r}, t) \tag{2.38}$$

where $\hat{H}$ is the *Hamiltonian operator* associated with the system total energy and, therefore, given by

$$\hat{H} = \hat{T} + \hat{V} \tag{2.39}$$

i.e., by the sum of the *kinetic energy operator* $\hat{T}$ and of the *potential energy operator* $\hat{V}$. Equation (2.38) is named the *Schrödinger equation*[6] and it is cast linear in order to fulfil the superposition principle given in equation (2.1). The derivation with respect to time is first order, so that just one boundary initial condition is needed and typically set as $\Psi(\mathbf{r}, t = 0) = f(x, y, z)$.

Let us suppose that $\Psi(\mathbf{r}, t)$ is the normalized solution of equation (2.38) over the volume Ω (see section 2.1.2). Then it holds that

$$i\hbar \frac{\partial \Psi}{\partial t} = \hat{H}\Psi \rightarrow i\hbar\Psi^* \frac{\partial \Psi}{\partial t} = \Psi^* \hat{H}\psi$$

$$-i\hbar \frac{\partial \Psi^*}{\partial t} = \hat{H}^* \Psi^* \rightarrow -i\hbar\Psi \frac{\partial \Psi^*}{\partial t} = \Psi \hat{H}^* \Psi^*$$

which by subtraction leads to

$$i\hbar \frac{\partial}{\partial t}(\Psi^* \Psi) = \Psi^* \hat{H}\Psi - \Psi \hat{H}^* \Psi^* \tag{2.40}$$

Since $\hat{H}$ is Hermitian, then

$$\frac{d}{dt} \int_\Omega \Psi^* \Psi \, d\mathbf{r} = 0 \tag{2.41}$$

or equivalently: *if the wavefunction $\Psi(\mathbf{r}, t)$ is normalized at time $t = 0$, then it will be so at the following time $t > 0$.* It is hard to underestimate the importance of this result: based on it, we could always use the particle wavefunction[7] to meaningfully predict the probability of finding it somewhere in the space, despite the fact it undergoes time evolution.

It is easy to prove that the particle probability density $|\Psi|^2 = \rho$ obeys the following *continuity equation*

$$\frac{\partial \rho}{\partial t} + \nabla \cdot \mathbf{j} = 0 \tag{2.42}$$

where

$$\mathbf{j} = \frac{\hbar}{2im}(\Psi^* \nabla \Psi - \Psi \nabla \Psi^*) \tag{2.43}$$

is the *probability density current vector* and, of course, m is the particle mass.

The solution of the Schrödinger equation is especially simple in the case of a *free particle* (i.e., $\hat{V} = 0$) in which case we get

$$\Psi(\mathbf{r}, t) = A \exp\left[i\left(\frac{\mathbf{p} \cdot \mathbf{r}}{\hbar} - \frac{Et}{\hbar} \right) \right] \tag{2.44}$$

---

[6] This is in honour of E Schrödinger who first introduced it in 1925–6
[7] Provided that initially normalized.

where $\mathbf{p} = \hbar\mathbf{k}$ is the momentum of the particle, $\mathbf{k} = \sqrt{2mE}/\hbar$ its wavevector (more precisely: the wavevector of the matter wave describing the particle), and E its energy. In equation (2.44) the prefactor A is the normalization constant.

2.4.2 Stationary states

Many quantum mechanical problems of interest in atomic and molecular systems are described by a *time-independent Hamiltonian operator*, as we will discuss in the next chapters. If this is the case, the differential operator appearing on the left or on the right hand side term of equation (2.38) only acts on time or on space, respectively. This makes it possible to separate the variables which Ψ depends on

$$\Psi(\mathbf{r}, t) = \psi(\mathbf{r})\xi(t) \qquad (2.45)$$

so that, by direct insertion into the Schrödinger equation, we eventually get

$$\xi(t) = \exp[-iEt/\hbar] \qquad (2.46)$$

while the equation left for $\psi(\mathbf{r})$ is found to be

$$\hat{H}\psi(\mathbf{r}) = E\psi(\mathbf{r}) \qquad (2.47)$$

that is called *eigenvalue equation for the Hamiltonian operator* or, alternatively, the *time-independent Schrödinger equation*[8]. Consistently with the picture we are elaborating, E is called *total energy* and it results independent of time.

The eigenstates with a well-defined energy are known as stationary states[9]. Their time evolution is described by the only function $\xi(t)$ since it holds equation (2.45), while it is straightforward to prove that for such stationary states the probability density ρ is constant as well (and, therefore, the space distribution of the stationary matter wave does not change in time). The same holds for its current vector $\mathbf{j}$.

The expectation value of an operator $\hat{F}$ is calculated as

$$\langle F \rangle = \int \Psi^*(\mathbf{r}, t)\hat{F}\Psi(\mathbf{r}, t)\, d\mathbf{r} = \int \psi^*(\mathbf{r})\hat{F}\psi(\mathbf{r})\, d\mathbf{r} \qquad (2.48)$$

and, therefore, it is constant in time for stationary states. Furthermore, if $[\hat{H}, \hat{F}] = 0$ then even the physical observable F has a well-defined value which is predicted by the same equation above. In summary, *in stationary states we have well-defined values either the energy and any other observable corresponding to an operator which commutes with the system Hamiltonian.*

---

[8] Sometimes, equation (2.47) is referred to simply as the Schrödinger equation. While this is in fact an abuse of notation, the context will help in understanding whether the equation actually refers to a time-dependent problem or it is rather addressed to calculate the energy eigenvalues.

[9] This wording is fully consistent with—in fact, inspired by—the definition of stationary state previously used in the Bohr model for the electronic structure of the hydrogen atom.

Finally, we observe an important mathematical property: if the potential V is everywhere finite, then the first derivative of the stationary wavefunction is a continuous function of the space variable $\mathbf{r} = (x, y, z)$. This implies that in any stationary problem where the potential energy undergoes a finite discontinuity at a given point, there both $\psi(\mathbf{r})$ and its derivative are continuous: this mathematical constraint provides the two boundary conditions that must be imposed to solve the solution of the Schrödinger equation.

2.4.3 Non-stationary states

It is plain that stationary states are not the only existing ones: there could in fact exist states $\phi(\mathbf{r}, t)$ not corresponding to any well-defined energy value. The most direct way to describe them and their properties is making use of stationary state wavefunctions. Let us suppose that the time-independent Hamiltonian operator $\hat{H}$ has a discrete and non-degenerate spectrum

$$\hat{H}\psi_n = E_n\psi_n \tag{2.49}$$

with $n = 1, 2, 3, \ldots$ so that its stationary state eigenfunctions form an orthonormal complete set. This implies that any other state vector $\phi(\mathbf{r}, t)$ (possibly non-stationary) can be written as the following linear combination

$$\phi(\mathbf{r}, t) = \sum_n a_n\psi_n(\mathbf{r})\exp[-iE_nt/\hbar] \tag{2.50}$$

where a_n are suitable constants. This result represents the most general way to describe *the time variation of the wavefunction describing an arbitrary quantum state*. Its mean energy (or, equivalently, the energy expectation value in the quantum state described by such a wavefunction) is given by

$$\langle E \rangle = \int \phi^*(\mathbf{r}, t)\hat{H}\phi(\mathbf{r}, t) \, d\mathbf{r} = \sum_n |a_n|^2 E_n \tag{2.51}$$

which, as expected, is not the energy of any stationary state but, interestingly enough, it does not depend on time. In contrast, the probability density of the state $\phi(\mathbf{r}, t)$ is calculated as

$$\rho(\mathbf{r}, t) = |\phi(\mathbf{r}, t)|^2 = \sum_{n,m} a_n^* a_m \psi_n^*(\mathbf{r})\psi_m(\mathbf{r})\exp[i(E_n - E_m)t/\hbar] \tag{2.52}$$

which demonstrates that the space distribution of a non-stationary matter wave undergoes time evolution.

2.4.4 Reconciling quantum and classical physics

At a glance, the picture emerging from the quantum theory developed so far could result somewhat disconcerting if compared to the classical description of physical phenomena. The two mostly disturbing features are likely (i) the probabilistic description[10] and (ii) the time evolution of physical observables[11]. It is worth attempting a reconciliation between quantum and classical pictures. To this aim we will first obtain an important formal result which will be next applied to derive an interesting and densely meaningful 'equation of motion'.

Let us consider an arbitrary non-stationary state ϕ and a physical quantity F associated with the quantum operator $\hat{F}$. Since we know that the expectation value $\langle f \rangle$ of the observable is

$$\langle f \rangle = \int \phi^* \hat{F} \phi \; d\mathbf{r} \tag{2.53}$$

we can calculate its time evolution as

$$\frac{d\langle f \rangle}{dt} = \int \left(\frac{\partial \phi^*}{\partial t} \hat{F} \phi + \phi^* \frac{\partial \hat{F}}{\partial t} \phi + \phi^* \hat{F} \frac{\partial \phi}{\partial t} \right) d\mathbf{r} \tag{2.54}$$

By means of the Schrödinger equation and taking profit of the Hermitian character of the Hamiltonian operator $\hat{H}$ we can write

$$\frac{d\langle f \rangle}{dt} = \int \phi^* \left(\frac{\partial \hat{F}}{\partial t} + \frac{1}{i\hbar} [\hat{F}, \hat{H}] \right) \phi \; d\mathbf{r} \tag{2.55}$$

which, for further convenience, can be cast in the form of an operator equation

$$\frac{d\hat{F}}{dt} = \frac{\partial \hat{F}}{\partial t} + \frac{1}{i\hbar} [\hat{F}, \hat{H}] \tag{2.56}$$

Interestingly enough, equation (2.55) implies that *if an operator does not explicitly depend on time and commutes with the Hamiltonian operator, then the corresponding observable has a constant expectation value* (in whatever state). Such an observable is named *quantum integral of motion*.

Let us now consider the quantum operators $\hat{r}_x$ e $\hat{p}_x$ associated with the position and linear momentum of a particle with mass m in one-dimensional motion along the x axis. Since both operators do not explicitly depend on time, from equation (2.56) we have

$$\frac{d\hat{r}_x}{dt} = \frac{1}{i\hbar} [\hat{r}_x, \hat{H}] \quad \frac{d\hat{p}_x}{dt} = \frac{1}{i\hbar} [\hat{p}_x, \hat{H}] \tag{2.57}$$

from which, by using the commuting rules given in equation (2.11), we get

$$\frac{d\hat{r}_x}{dt} = \frac{1}{m} \hat{p}_x \quad \frac{d\hat{p}_x}{dt} = -\frac{\partial \hat{V}(x)}{\partial x} \tag{2.58}$$

---

[10] According to quantum theory, a particle is never 'here' or 'there' but, rather, it has a given probability of being anywhere within the allowed space.

[11] The Schrödinger equation makes a clear distinction between stationary and non-stationary states as for the time evolution of observables which, in some cases, could not have well-defined values but, rather, only expectation ones.

or

$$m\frac{d^2\hat{r}_x}{dt^2} = -\frac{\partial\hat{V}}{\partial x}$$ (2.59)

where we have made use of the expression $\hat{H} = -(\hbar^2/2m)\partial^2/\partial x^2 + \hat{V}(x)$ for the Hamiltonian operator. This result, known as *Ehrenfest theorem*, establishes a clean and substantial conceptual link between the quantum mechanical theory based on wavefunctions, operators, and expectation values and the classical Newton equation of motion. This is clearly obtained by switching from equation (2.59) to its corresponding version using expectation values

$$m\frac{d^2}{dt^2}\int \phi^*\hat{r}_x\phi \, d\mathbf{r} = -\int \phi^*\frac{\partial\hat{V}}{\partial x}\phi \, d\mathbf{r}$$ (2.60)

which states that *the expectation value of the particle position follows a trajectory imposed by the expectation value of the force* (here identified with the expectation value of the potential derivative with respect to the space coordinate). In other words, if we imagine that the mass of the particle is concentrated in the position provided by the expectation value of the $\hat{r}_x$ operator, then its dynamics is Newtonian.

2.5 Systems of identical particles

Let us now move to consider a system of *identical particles*. In quantum mechanics, such identical objects are also *indistinguishable*. This is indeed a subtle new concept with many deep conceptual implications, setting another sharp difference between quantum and classical descriptions. While a full formal treatment of quantum indistinguishability can be found elsewhere [1, 3, 4], here it is enough to develop a phenomenological argument to explain it. Let us preliminarily clarify that two particles are identical provided that they have the same characteristics (i.e., same intrinsic physical properties like, for instance, the mass, the charge, and so on). Identical classical particles can be nevertheless distinguished by taking into consideration their position. This is, for instance, the case of a set of billiard balls: even if we assume that they all have same shape, dimension, mass, color, ... we can nevertheless state that each one occupies a well-defined position on the billiard table. This allows us to unambiguously attach to any ball a label, say a number, to identify it: in short, billiard balls are identical, but distinguishable. Let us now consider a set of quantum particles like, e.g., electrons: while they do have the same intrinsic physical properties (same charge, same mass, same magnetic dipole moment[12]), we cannot assign to them a well-defined position, but only an occupation probability for any point in the space. This in fact prevents distinguishing them: in short, electrons are identical and indistinguishable.

Let us then consider a set of *N identical and indistinguishable particles* of mass *m* described by the Hamiltonian operator

---

[12] The electron magnetic moment is related to an intrinsic degree of freedom named spin. Its physics will be extensively discussed in the next chapters.

$$\hat{H} = -\frac{\hbar^2}{2m}\sum_{i=1}^{N}\nabla_i^2 + \hat{V}(1, 2,\dots, \mu,\dots, \nu,\dots, N-1, N) \tag{2.61}$$

where $\hat{V}(1, 2,\dots, \mu,\dots, \nu,\dots, N-1, N)$ is the potential energy operator which, in general, depends on the coordinates of all particles which hereafter will be written in compact notation $\mathbf{r}_1 = 1, \mathbf{r}_2 = 2,\dots, \mathbf{r}_N = N$. It is plain to understand that, just because of particle identity and indistinguishability, $\hat{H}$ *must be invariant upon interchanging any two particles.* In other words, the index swapping $\mu \leftrightarrow \nu$ does not change $\hat{H}$ and the system remains just the same. However, we remark that, in principle, the same operation could affect the eigenfunction $\phi(\mathbf{r}_1, \mathbf{r}_2,\dots, \mathbf{r}_N) = \phi(1, 2,\dots, N)$ of $\hat{H}$ describing the state of the system. This possibility must be duly explored in detail.

2.5.1 Wavefunction symmetry

Let $\hat{P}_{\mu\leftrightarrow\nu}$ be the *permutation operator* which acts on the eigenfunctions of $\hat{H}$ by swapping the indices $\mu \leftrightarrow \nu$. It is easy to prove that

$$[\hat{H}, \hat{P}_{\mu\leftrightarrow\nu}] = 0 \tag{2.62}$$

and, therefore, *ϕ is a wavefunction of either $\hat{H}$ and $\hat{P}_{\mu\leftrightarrow\nu}$.* The eigenvalue equation for the permutation operator reads as

$$\hat{P}_{\mu\leftrightarrow\nu}\phi(1, 2,\dots, \mu,\dots, \nu,\dots, N-1, N)$$
$$= p\,\phi(1, 2,\dots, \mu,\dots, \nu,\dots, N-1, N) \tag{2.63}$$

where p is a real number since $\hat{P}_{\mu\leftrightarrow\nu}$ is Hermitian. By swapping the same indices a second time we get

$$\hat{P}_{\mu\leftrightarrow\nu}^2\phi(1, 2,\dots, \mu,\dots, \nu,\dots, N-1, N)$$
$$= p^2\phi(1, 2,\dots, \mu,\dots, \nu,\dots, N-1, N) \tag{2.64}$$

and since it holds that

$$\hat{P}_{\mu\leftrightarrow\nu}\phi(1, 2,\dots, \mu,\dots, \nu,\dots, N-1, N)$$
$$= \phi(1, 2,\dots, \nu,\dots, \mu,\dots, N-1, N) \tag{2.65}$$

we eventually obtain

$$\hat{P}_{\mu\leftrightarrow\nu}^2\phi(1, 2,\dots, \mu,\dots, \nu,\dots, N-1, N)$$
$$= \phi(1, 2,\dots, \mu,\dots, \nu,\dots, N-1, N) \tag{2.66}$$

which dictates $p = \pm 1$. This result is of paramount importance: we have proved that *any wavefunction describing a set of identical and indistinguishable particles is either symmetric or antisymmetric under particle interchange*:

$$\hat{P}_{\mu\leftrightarrow\nu}\phi = +\phi \quad \text{symmetric wavefunction} \tag{2.67}$$

$$\hat{P}_{\mu\leftrightarrow\nu}\phi = -\phi \quad \text{antisymmetric wavefunction} \tag{2.68}$$

The symmetry property of a wavefunction cannot be altered by any external action and it therefore represents an intrinsic property of the specific set of particles considered. In particular, identical and indistinguishable particles are said *fermions* (*bosons*) if their collective quantum states are described by antisymmetric (symmetric) wavefunctions [2].

The rationalization of a large number of different experimental investigations drives to the conclusion that *any elementary particle existing in Nature is either a fermion or a boson*. More specifically, with reference to the limited set of elementary particles introduced in chapter 1, it is found that: *electrons, protons, and neutrons are fermions*, while *photons are bosons*. The fermion-/boson-like character of an elementary particle is ultimately determined by its *intrinsic degree of freedom* named *spin*. While in the following chapters the electron spin will be introduced and described based on experimental findings, we remark that the physical origin of the particle spin can only be accounted for by a fully relativistic quantum theory, as well as the link between such a spin and symmetry property of corresponding wavefunction [5].

2.5.2 Pauli principle

The antisymmetric character of the fermion wavefunction[13] has a very important consequence which we will derive under the assumption that a system of N fermions is subjected to the action of a potential described by the operator

$$\hat{V}(1, 2,\dots, N) = \sum_{i=1}^{N}\hat{v}_i \tag{2.69}$$

where $\hat{v}_i$ is a potential energy operator just acting on the ith particle (i.e., it only depends on the coordinates of the ith atom). Interestingly enough, equation (2.69) corresponds to the central-field approximation largely adopted to treat multi-electron atoms (see chapter 5). Since the additivity of $\hat{V}$, the original many-body problem

$$\hat{H}\Phi(1, 2,\dots, N) = E\Phi(1, 2,\dots, N) \tag{2.70}$$

where $\Phi(1, 2,\dots, N)$ is the total wavefunction (depending on the coordinates of all particles) and $\hat{H}$ and E are the system Hamiltonian and energy, respectively, is reduced to N different single-particle problems

$$\hat{h}_i\varphi_{n_i}(i) = \varepsilon_i\varphi_{n_i}(i) \quad i = 1, 2,\dots, N \tag{2.71}$$

---

[13] This is for us the mostly interesting case, since electrons are fermions and their physics underlines most of atomic and molecular physics.

where it holds that

$$\hat{H} = \sum_{i=1}^{N} \hat{h}_i = \sum_{i=1}^{N} \left(-\frac{\hbar^2}{2m} \nabla_i^2 + \hat{v}_i \right) \tag{2.72}$$

In equation (2.71) we have indicated by n_i a suitable set of quantum numbers[14] fully characterizing the state vector $\varphi_{n_i}(i)$ of the ith fermion with energy ε_i. Of course we have $E = \sum_i \varepsilon_i$.

Because of indistiguishability, the eigenfunction of the full quantum problem given in equation (2.70) can be cast in the form

$$\Phi(1, 2, \ldots, N) = \frac{1}{\sqrt{N!}} \sum_{\xi} (-1)^{\xi} \hat{P}_{\xi} \varphi_{n_1}(1) \varphi_{n_2}(2) \cdots \varphi_{n_N}(N)$$

$$= \frac{1}{\sqrt{N!}} \begin{vmatrix} \varphi_{n_1}(1) & \varphi_{n_1}(2) & \cdots & \varphi_{n_1}(N) \\ \varphi_{n_2}(1) & \varphi_{n_2}(2) & \cdots & \varphi_{n_2}(N) \\ \cdots & \cdots & \cdots & \cdots \\ \varphi_{n_N}(1) & \varphi_{n_N}(2) & \cdots & \varphi_{n_N}(N) \end{vmatrix} \tag{2.73}$$

where $\hat{P}_{\xi}$ is the operator corresponding to ξ consecutive permutations of particle pairs. The equation (2.73) is named the *Slater determinant*, representing the most general form of the wavefunction of a system of identical and indistinguishable fermions. *This form of* Φ *naturally embodies the antisymmetric character of the state function*, since by swapping two columns the Slater determinant will change sign. The physical motivation behind this mathematical formulation of Φ is subtle: since particles cannot be distinguished, we cannot think to 'take' a given particle and 'place' it on a specific state; therefore, we must write Φ in the only form where *any possible combination particle* $\leftrightarrow$ *state is considered explicitly*. This results in the determinant form.

The Slater determinant has another very important mathematical property, namely: *it vanishes if two rows are equal*. Let us suppose this is the case for row ith and row jth: this implies that $n_i = n_j$, indicating that two fermions have exactly the same set of quantum numbers. We conclude by saying that *a system of identical and indistinguishable fermions cannot occupy a many-body state where two single-particle states are equal*. This statement is commonly referred to as *Pauli principle* and can be equivalently rephrased as: *in a system of identical and indistinguishable fermions it is impossible to have two particles with the same set of quantum numbers*. Pauli principle has many important implications in atomic physics, underlying the actual electronic structure of multi-electrons as well as the ordering criteria

---

[14] The concept of quantum number was previously introduced in equation (2.14) for a given operator. In general, a single-particle wavefunction φ can simultaneously be an eigenfunction of two or more operators, the spectrum of each being charaterized by a specific quantum number. This means that, in general, φ is associated with two or more quantum numbers: in equation (2.71) the symbol n_i indicates the full set of such quantum numbers.

providing the periodic table of elements. They will be extensively addressed in the next chapters.

2.6 Matrix notation

Let us consider the eigenvalue problem

$$\hat{F}\varphi_f = f\varphi_f \tag{2.74}$$

where $\hat{F}$ is an Hermitian operator with a discrete and non-degenerate spectrum. By making use of the eigenfunctions ψ_n of the Hamiltonian operator, we can write

$$\varphi_f = \sum_n a_n^{(f)}\psi_n \tag{2.75}$$

so that

$$\hat{F}\left(\sum_n a_n^{(f)}\psi_n\right) = f\left(\sum_n a_n^{(f)}\psi_n\right)$$

$$\sum_n (\hat{F}\psi_n - f\psi_n)a_n^{(f)} = 0$$

If we now multiply on the left by ψ_m^* (with $m \neq n$) and then integrate over all available space, we get

$$\sum_n \left(\int \psi_m^* \hat{F}\psi_n \, d\mathbf{r} - f \int \psi_m^* \psi_n \, d\mathbf{r}\right)a_n^{(f)} = 0 \tag{2.76}$$

By further setting

$$\int \psi_m^* \hat{F}\psi_n \, d\mathbf{r} = F_{mn} \tag{2.77}$$

we define the *the matrix elements of the operator $\hat{F}$ on the basis set of the eigenfunctions of the Hamiltonian operator*. This result reflects a more general feature: whenever we benefit from a complete and orthonormal set of state vectors, any operator can be represented as a matrix, whose entries are given by integrals as in equation (2.77). Within such a matrix representation the eigenvalue problem cast in equation (2.74) is replaced by the algebraic system

$$\sum_n (F_{mn} - f\delta_{mn})a_n^{(f)} = 0 \tag{2.78}$$

which admits non-trivial solutions provided that

$$\det|F_{mn} - f\delta_{mn}| = 0 \tag{2.79}$$

In summary, *the fundamental quantum problem given by equation (2.74) is recast in the form of a matrix diagonalization*: the eigenvalues are all the possible values f of the physical observable associated with the quantum operator $\hat{F}$, while the column eigenvectors provide the coefficients of the linear combination given in equation (2.75) and allow one to calculate the corresponding eigenfunctions φ_f.

It is worth mentioning that quantum mechanics is often formulated according to the Dirac 'bra'-'ket' notation. While we will not make use of such symbols, the interested reader could refer to the original Dirac textbook for further details [6].

2.7 Perturbation theory

2.7.1 The concept of 'perturbation'

Let us consider an initially isolated physical system, whose quantum features are known through the solution of the equation

$$\hat{H}_0 \psi_n^{(0)} = E_n^{(0)} \psi_n^{(0)} \tag{2.80}$$

where, once again, we assume a discrete and non-degenerate spectrum. Whenever we apply some external field on it we say that *we are applying a perturbation*, understanding that the *'perturbation' is a disturbance of the initial unperturbed state described by* $\hat{H}_0$.

If we can describe the physical perturbation by a suitable operator $\hat{H}_{\text{pert}}$, then the new quantum problem is described by

$$\left(\hat{H}_0 + \hat{H}_{\text{pert}} \right) \phi = E\phi \tag{2.81}$$

where new eigenfunctions ϕ and new energies E appear. The theory of perturbations aims at elaborating general strategies to solve such an equation. It is suitable to write the perturbation operator as

$$\hat{H}_{\text{pert}} = \lambda \hat{W} \tag{2.82}$$

where λ is a dimensionless real coupling parameter describing the strength of the perturbation: while for small λ values the situations described by equations (2.80) and (2.81) are very similar, a large λ value will identify very different, respectively unperturbed and perturbed, physically situations. In most applications we will come across the perturbation is small, so that the perturbed quantum states do not differ from unperturbed ones to a large extent. This corresponds to a small perturbation strength with respect to the energy spectrum of the unperturbed system. In these situations, we can formally proceed by expanding the perturbation operator in powers of λ, limiting to the first terms.

2.7.2 Time-independent perturbations

Let us consider a *time-independent small perturbation* $\hat{H}_{\text{pert}} = \lambda \hat{W}$ and use the unperturbed wavefunctions $\psi_n^{(0)}$ as a basis set to represent the perturbed state vector ϕ. We accordingly get

$$(\hat{H}_0 + \lambda\hat{W})\sum_n a_n \psi_n^{(0)} = E\sum_n a_n \psi_n^{(0)}$$

$$\sum_n a_n\left(\hat{H}_0\psi_n^{(0)} + \lambda\hat{W}\psi_n^{(0)}\right) = E\sum_n a_n\psi_n^{(0)} \tag{2.83}$$

$$\sum_n a_n(E - E_n^{(0)})\psi_n^{(0)} = \lambda\sum_n a_n\hat{W}\psi_n^{(0)}$$

By multiplying on the left by $\left[\psi_m^{(0)}\right]^*$ and integrating over the available space we obtain

$$(E - E_m^{(0)})a_m = \lambda\sum_n a_n W_{mn} \tag{2.84}$$

where, according to equation (2.77), W_{mn} is the *matrix element of the perturbation operator on the basis set of the unperturbed wavefunctions.*

Formally, equation (2.84) provides the energy spectrum $E = E_l$ (with $l = 1, 2, 3,...$) of the perturbed system. In order to proceed, we exploit the assumed condition of *small perturbation*: the perturbed energies E_l will only slightly differ from the unperturbed ones and therefore

$$E_l = E_l^{(0)} + \lambda E_l^{(1)} + \lambda^2 E_l^{(2)} + \cdots \tag{2.85}$$

where $E_l^{(1)}$, $E_l^{(2)}$,... define, respectively, the *first-order, second-order, ... correction* to the unperturbed energy $E_l^{(0)}$. Similarly, the coefficients a_n of the linear combination $\phi = \sum_n a_n\psi_n^{(0)}$ are set as

$$a_n = \delta_{nl} + \lambda a_n^{(1)} + \lambda^2 a_n^{(2)} + \cdots \tag{2.86}$$

By inserting equation (2.85) and equation (4.8) into equation (2.84) we easily obtain

$$[(E_l^{(0)} - E_m^{(0)}) + \lambda E_l^{(1)} + \lambda^2 E_l^{(2)} + \cdots](\delta_{ml} + \lambda a_m^{(1)} + \lambda^2 a_m^{(2)} + \cdots)$$
$$= \lambda\sum_n W_{mn}(\delta_{nl} + \lambda a_n^{(1)} + \lambda^2 a_n^{(2)} + \cdots) \tag{2.87}$$

We must distinguish two cases:

1. $\boxed{m \neq l}$ By matching terms of the same order in λ we get

$$a_m^{(1)}\left(E_l^{(0)} - E_m^{(0)}\right) = \sum_n W_{mn}\delta_{nl} = W_{ml} \tag{2.88}$$

$$a_m^{(2)}\left(E_l^{(0)} - E_m^{(0)}\right) + a_m^{(1)}E_l^{(1)} = \sum_n W_{mn}a_n^{(1)} \tag{2.89}$$

$$\cdots = \cdots$$

and obtain a very useful expression that we will largely use in the following

$$a_m^{(1)} = \frac{W_{ml}}{E_l^{(0)} - E_m^{(0)}} \tag{2.90}$$

2. $\boxed{m = l}$ By matching terms of the same order in λ we get

$$E_l^{(1)} = \sum_n W_{mn}\delta_{nl} = \sum_n W_{ln}\delta_{nl} = W_{ll} \tag{2.91}$$

$$a_l^{(1)}E_l^{(1)} + E_l^{(2)} = \sum_n W_{mn}a_n^{(1)} \tag{2.92}$$
$$\cdots = \cdots$$

The above equations are useful in calculating the first-order correction to energy

$$\begin{aligned}
E_l &= E_l^{(0)} + \lambda E_l^{(1)} \\
&= E_l^{(0)} + \lambda W_{ll} \\
&= E_l^{(0)} + \int \left[\psi_l^{(0)}\right]^* (\lambda \hat{W})\psi_l^{(0)} \, d\mathbf{r} \\
&= E_l^{(0)} + H_{\text{pert},ll}
\end{aligned} \tag{2.93}$$

which states that *in first approximation the correction to the energy values is provided by the expectation value of the perturbation operator on the basis of the unperturbed wavefunctions*. The next-order correction to energy is given by

$$a_l^{(1)}E_l^{(1)} + E_l^{(2)} = W_{ll}a_l^{(1)} + \sum_{n \neq l} W_{mn}a_n^{(1)}$$

$$E_l^{(2)} = \sum_{n \neq l} \frac{W_{ln}W_{nl}}{E_l^{(0)} - E_n^{(0)}} = \sum_{n \neq l} \frac{|W_{nl}|^2}{E_l^{(0)} - E_n^{(0)}} \tag{2.94}$$

so that

$$E_l = E_l^{(0)} + \lambda W_{ll} + \lambda^2 \sum_{n \neq l} \frac{|W_{nl}|^2}{E_l^{(0)} - E_n^{(0)}} \tag{2.95}$$

or equivalently

$$E_l = E_l^{(0)} + H_{\text{pert},ll} + \sum_{n \neq l} \frac{|H_{\text{pert},nl}|^2}{E_l^{(0)} - E_n^{(0)}} \tag{2.96}$$

which proves that *to the second order in the perturbation, energy corrections quadratically depend on the off-diagonal matrix elements of the perturbation operator.*

Let us now turn to eigenfunctions ϕ_l corresponding to energies E_l and focus just on first-order corrections

$$\phi_l = \sum_n (\delta_{nl} + \lambda a_n^{(1)})\psi_n^{(0)} \tag{2.97}$$

$$= \sum_n \delta_{nl}\psi_n^{(0)} + \lambda a_l^{(1)}\psi_l^{(0)} + \lambda \sum_{n \neq l} a_n^{(1)}\psi_n^{(0)}$$

$$= \psi_l^{(0)} + \lambda a_l^{(1)}\psi_l^{(0)} + \lambda \sum_{n \neq l} \frac{W_{nl}}{E_l^{(0)} - E_n^{(0)}}\psi_n^{(0)} \qquad (2.98)$$

$$= \psi_l^{(0)} + \lambda a_l^{(1)}\psi_l^{(0)} + \sum_{n \neq l} \frac{H_{\text{pert},nl}}{E_l^{(0)} - E_n^{(0)}}\psi_n^{(0)}$$

In order to determine the term $\lambda a_l^{(1)}$, we can compute the product $\phi_l^*\phi_l$ to the first order in λ by imposing the normalization condition $\int \phi_l^*\phi_l \, d\mathbf{r} = 1$: this leads to $a_l^{(1)} + [a_l^{(1)}]^* = 0$ which, in turn, means that the coefficients $a_l^{(1)}$ are imaginary numbers. Since eigenfunctions are defined to within an arbitrary phase factor, as discussed in section 2.1.2, we can conveniently put $a_l^{(1)} = 0$ for any index l. In conclusion, the first-order perturbed wavefunctions are obtained as

$$\phi_l = \psi_l^{(0)} + \sum_{n \neq l} \frac{H_{\text{pert},nl}}{E_l^{(0)} - E_n^{(0)}}\psi_n^{(0)} \qquad (2.99)$$

We can now better clarify the condition under which the theory of time-independent perturbations here outlined can be applied: the above corrections to both energies and wavefunctions can be safely used provided that $|H_{\text{pert},nl}| < |E_l^{(0)} - E_n^{(0)}|$. As anticipated, this corresponds to a physical situation in which the energy content of the perturbation is small compared to the differences between unperturbed states.

2.7.3 Time-dependent perturbations

The action of a perturbation varying in time represents quite a few circumstances typically found in atomic and molecular physics, significantly including the action of an electromagnetic field. Let us then consider a physical system which, over a time lapse τ, is subjected to a *time-dependent perturbation* described by the perturbation operator

$$\hat{H}_{\text{pert}}(t) = \begin{cases} \hat{W}(t) & \text{for } 0 \leqslant t \leqslant \tau \\ 0 & \text{for } t < 0 \text{ and } t > \tau \end{cases} \qquad (2.100)$$

which corresponds to the Schrödinger problem

$$i\hbar \frac{\partial \phi(\mathbf{r}, t)}{\partial t} = [\hat{H}_0 + \hat{W}(t)]\phi(\mathbf{r}, t) \qquad (2.101)$$

where, as above, we have indicated by $\hat{H}_0$ the unperturbed Hamiltonian operator with eigenfunctions $\psi_l^{(0)}$ and eigenvectors $E_l^{(0)}$ while, as discussed in section 2.4.3, we can write

$$\phi(\mathbf{r}, t) = \sum_l a_l(t)\psi_l^{(0)}(\mathbf{r})\exp[-iE_l^{(0)}t/\hbar] \qquad (2.102)$$

We observe that *before the perturbation takes place* the system necessarily occupies an eigenstate, say $\psi_m^{(0)}$, of the unperturbed Hamiltonian

$$\phi_{\text{before}}(\mathbf{r}, t) = \psi_m^{(0)}(\mathbf{r})\exp[-iE_m^{(0)}t/\hbar] \text{ for } t < 0 \tag{2.103}$$

while *at the end of the perturbation action* it will occupy a generic state

$$\phi_{\text{after}}(\mathbf{r}, t) = \sum_n a_{mn}(\tau)\psi_n^{(0)}(\mathbf{r})\exp[-iE_n^{(0)}t/\hbar] \text{ for } t > \tau \tag{2.104}$$

where the coefficients $a_{mn}(\tau)$ depend either on the state initially occupied by the system and by the time duration of the perturbation. At any time $t > \tau$, the probability that the system will occupy a state with energy $E_n^{(0)}$ is given by $|a_{mn}(\tau)|^2$ and, therefore, we state that

$$\mathcal{P}_{m \to n} = |a_{mn}(\tau)|^2 \tag{2.105}$$

represents *the probability that, under the action of the perturbation $\hat{H}_{\text{pert}}(t)$, the system undergoes a transition from the initial state $\psi_m^{(0)}$ with energy $E_m^{(0)}$ to the final state $\psi_n^{(0)}$ with energy $E_n^{(0)}$ in the time interval τ*. The key goal of time-dependent perturbation theory is calculating such transition probability $\mathcal{P}_{m \to n}$.

By inserting equation (2.102) into equation (2.101), further multiplying on the left by $\left[\psi_n^{(0)}\right]^*$, and eventually integrating over the whole available space we get

$$i\hbar\frac{da_n(t)}{dt} = \sum_l a_l(t)W_{nl}\exp[i\omega_{nl}t] \tag{2.106}$$

where $\hbar\omega_{nl} = E_n^{(0)} - E_l^{(0)}$. If the perturbation is small and it acts over a suitably short time lapse, then a first-order solution is obtained by approximating $a_l(t) \sim a_l(0) = \delta_{lm}$

$$i\hbar\frac{da_{nm}(t)}{dt} = W_{nm}\exp[i\omega_{nm}t] \tag{2.107}$$

where on the left hand site we have set $a_n(t) = a_{nm}(t)$ to emphasize the very fact that the system initially occupies the state $\psi_m^{(0)}$. The first-order expression for the transition probability is immediately obtained as

$$\mathcal{P}_{m \to n} = |a_{mn}(\tau)|^2 = \frac{1}{\hbar^2}\left|\int_0^\tau W_{nm}\exp[i\omega_{nm}t]\,dt\right|^2 \tag{2.108}$$

which can easily be worked out in two interesting cases, namely: when the perturbation is either constant or harmonic over the time interval τ.

If $\hat{W}(t) = \hat{W} = $ constant for $0 \leqslant t \leqslant \tau$, then W_{nm} in equation (2.108) does not depend on time and therefore[15]

---

[15] We remark that this result is formally obtained under the combined condition that τ is sufficiently long to replace equation (2.98) with equation (2.99), but still shorter than the lifetime of the initial state [2, 3].

$$\mathcal{P}_{m \to n} = 2 \, |W_{nm}|^2 \, \frac{1 - \cos\left[\dfrac{\tau}{\hbar}(E_n^{(0)} - E_m^{(0)})\right]}{(E_n^{(0)} - E_m^{(0)})^2} \tag{2.109}$$

which is transformed into

$$\mathcal{P}_{m \to n} = \frac{2\pi}{\hbar}\tau \, |W_{nm}|^2 \, \delta(E_n^{(0)} - E_m^{(0)}) \tag{2.110}$$

Since, under the above assumption, the transition probability linearly grows in time, it is conveniently defined by the *transition probability per unit time*

$$\frac{\mathcal{P}_{m \to n}}{\tau} = \frac{2\pi}{\hbar} \, |W_{nm}|^2 \, \delta(E_n^{(0)} - E_m^{(0)}) \tag{2.111}$$

We now consider the much more interesting case when $\hat{W}(t)$ displays a harmonic time dependence

$$\hat{W}(t) = \hat{w}\exp[\pm i\omega t] \text{ for } 0 \leqslant t \leqslant \tau \tag{2.112}$$

Following the same procedure as before, we get

$$\mathcal{P}_{m \to n} = \frac{2\pi}{\hbar}\tau \, |w_{nm}|^2 \, \delta(E_n^{(0)} - E_m^{(0)} \pm \hbar\omega) \tag{2.113}$$

where the corresponding *transition probability per unit time* is calculated to be

$$\frac{\mathcal{P}_{m \to n}}{\tau} = \frac{2\pi}{\hbar} \, |w_{nm}|^2 \, \delta(E_n^{(0)} - E_m^{(0)} \pm \hbar\omega) \tag{2.114}$$

This remarkable result is known as the *Fermi golden rule*: it dictates that under the action of a harmonic perturbation:
1. the transition probability per unit time is proportional to the square matrix element of the perturbation operator, calculated for the two states involved in the transition process;
2. the transition does conserve energy or, equivalently, the energy difference between the initial and final state matches a quantum of energy of the perturbation field.

In the case where the perturbation corresponds to a harmonic electromagnetic field, the Fermi golden rules states that a photon is absorbed if $E_n^{(0)} > E_m^{(0)}$, while a photon is emitted in the opposite case $E_n^{(0)} < E_m^{(0)}$. Interesting enough, since we have assumed that the perturbation is small (i.e., the intensity of the electromagnetic field is weak), the unperturbed energy levels still represent physical states of the system under the action of the perturbation, whose only effect is, therefore, just to promote transitions by emission or absorption of photons.

Further reading and references

[1] Sakurai J J and Napolitano J 2011 *Modern Quantum Mechanics* 2nd edn (Reading, MA: Addison-Wesley)

[2] Miller D A B 2008 *Quantum Mechanics for Scientists and Engineers* (New York: Cambridge University Press)

[3] Griffiths D J and Schroeter D F 2018 *Introduction to Quantum Mechanics* 3rd edn (Cambridge: Cambridge University Press)

[4] Bransden B H and Joachain C J 2000 *Quantum Mechanics* (Englewood Cliffs, NJ: Prentice-Hall)

[5] Greiner W 1990 *Relativistic Quantum Mechanics* (Heidelberg: Springer)

[6] Dirac P A M 1958 *The Principles of Quantum Mechanics* (Oxford: Oxford University Press)

Part II

Atomic physics

IOP Publishing

Atomic and Molecular Physics (Second Edition)
A primer
Luciano Colombo

Chapter 3

One-electron atoms

Syllabus—*We begin our study of atomic physics by relying on the phenomenological notions summarized in chapter 1 and by fully exploiting the quantum mechanical formalism outlined in chapter 2. A hierarchy of approximations will be introduced in the first instance to describe the hydrogen atom, which allows one to focus on the main features of its energy spectrum and quantum properties, including the key 'atomic orbital' concept. Next, these results will be extended to the more general case of hydrogenic atoms. The physical picture will be made more rigorous by introducing the magnetic interactions, included those linked to the intrinsic spin degree of freedom, leading to the full description of the fine and hyperfine structure of one-electron atoms. The effects of a static external magnetic or electric field will be eventually discussed by calculating the corresponding energy shifts and splittings of the energy levels.*

3.1 The hydrogen atom

3.1.1 Problem definition and some useful approximations

The most simple of all atoms is *hydrogen*, consisting of a nucleus with mass number $A = 1$ and atomic number $Z = 1$, around which a single electron moves[1]. This elementary atomic system is therefore described as *a bound electron–proton pair*. We will start our journey from this case because either it paradigmatically contains some of the most relevant features in atomic physics and it represents an atomic system of paramount importance in Nature.

If we attribute a mass m_p to the proton and a mass m_e to the electron, then the classical total energy E_{tot} of a hydrogen atom is written as

$$E_{tot} = \frac{P_{cm}^2}{2(m_p + m_e)} + \frac{p^2}{2m_{eff}} - \frac{1}{4\pi\varepsilon_0}\frac{e^2}{r} \qquad (3.1)$$

---

[1] We have obviously adopted the Rutherford planetary model.

doi:10.1088/978-0-7503-5734-0ch3

where the first term describes the translational motion of the atomic system through the linear momentum $\mathbf{P}_{cm}$ of its centre-of-mass, while the second term accounts for the relative electron–proton motion in terms of an effective particle with mass $1/m_{eff} = 1/m_p + 1/m_e$ and momentum $\mathbf{p}$. The last contribution describes the Coulomb interaction in terms of the nuclear and electronic charges, respectively $\pm e$, and their relative distance r.

In order to proceed with the setup of the quantum problem, we preliminarily observe that the centre-of-mass motion in fact is not an atomic physics problem: accordingly, we will assume considering *the atom at rest in an inertial frame of reference*. Furthermore, we will conveniently approximate the situation $m_p \gg m_e$ by assuming *an infinite nuclear mass*. Finally, we will disregard any relativistic effect and, therefore, the only mass still surviving in the problem definition is the *electron rest mass* m_e.

Another more subtle approximation we implicitly adopted in writing equation (3.1) consists in *neglecting any magnetic interaction*. As a matter of fact, an electron moving around a proton along a close orbit does correspond to a current which, in turn, is equivalent to a magnetic dipole according to Ampère's principle of equivalence [1, 2]. Nevertheless, in equation (3.1) no evidence of related magnetic interactions is reported[2]. While this missing information makes our view of the problem surely incomplete, it is heuristically motivated by the fact that magnetic interactions are much weaker than Coulomb ones (more specifically, we will provide evidence that they are 10^3–10^4 times smaller than electrostatic coupling) and therefore marginal in the first instance.

All the above approximations will be later readdressed and in most cases released. Here what we aim at is basically *working out a first quantum picture of the hydrogen atom*, without treating it in full detail and thus avoiding overwhelming mathematics. Interesting enough, we will prove that the resulting (approximated) picture is suitable to explain some important experimental findings and to shed light on the very physics of a hydrogen atom.

The reformulated classical total energy

$$E_{tot} = \frac{p^2}{2m_e} - \frac{1}{4\pi\varepsilon_0} \frac{e^2}{r} \tag{3.2}$$

(where $\mathbf{p}$ is now the electron linear momentum) is easily translated into a *Hamiltonian operator for the hydrogen atom*

$$\hat{H} = -\frac{\hbar^2}{2m_e}\nabla^2 - \frac{1}{4\pi\varepsilon_0}\frac{e^2}{r} = -\frac{\hbar^2}{2m_e}\nabla^2 + \hat{V}(r) \tag{3.3}$$

---

[2] As explained next in this chapter, electrons have an additional intrinsic degree of freedom, named spin, which is associated with additional non-orbital magnetic interactions. Even this contribution is not accounted for in equation (3.1): a full treatment of spin-related features will be later developed when elaborating a more sophisticated picture.

by following the recipes described in chapter 2. In this equation it is used the compact notation $V(r) = -e^2/4\pi\varepsilon_0 r$ and it is understood that

$$\nabla^2 = \frac{\partial^2}{\partial x^2} + \frac{\partial^2}{\partial y^2} + \frac{\partial^2}{\partial z^2} \quad \text{and} \quad r = \sqrt{x^2 + y^2 + z^2} \tag{3.4}$$

with respect to a Cartesian frame of reference centred on the proton, as shown in figure 3.1.

The Hamiltonian operator appearing in equation (3.3) does not depend on time and, therefore, as shown in section 2.4.2 we can separate the space and time coordinates which the *total electron wavefunction* $\Psi(\mathbf{r}, t)$ depends on

$$\Psi(\mathbf{r}, t) = \psi(\mathbf{r})\exp[-iEt/\hbar] \tag{3.5}$$

with

$$\hat{H}\psi(\mathbf{r}) = E\psi(\mathbf{r}) \tag{3.6}$$

We have naturally introduced the *electron energy* E and the wavefunction $\psi(\mathbf{r})$ describing the corresponding *stationary state*. The final goal is solving equation (3.6). While it is in principle possible to do that in Cartesian coordinates, it is much more convenient—in fact, much more meaningful—to fully exploit the *spherical symmetry of the problem* which is dictated by the fact that the Coulomb potential $V(r)$ is central. Therefore, with reference to figure 3.1, we introduce a set of polar coordinates such that

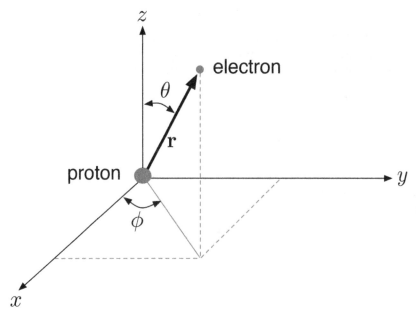

Figure 3.1. The Cartesian frame of reference centred on the proton and the corresponding set of polar coordinates for the electron position.

$$\psi(\mathbf{r}) = \psi(x, y, z) \quad \rightarrow \quad \psi(r, \theta, \phi) \tag{3.7}$$

and recast the Hamiltonian operator in the form[3]

$$\hat{H} = -\frac{\hbar^2}{2m_e} \frac{1}{r^2} \frac{\partial}{\partial r}\left(r^2 \frac{\partial}{\partial r}\right) + \frac{1}{2m_e r^2}\hat{L}^2 + \hat{V}(r) \tag{3.8}$$

where we have introduced the polar representation of the *square angular momentum operator*

$$\hat{L}^2 = -\hbar^2\left[\frac{1}{\sin\theta}\frac{\partial}{\partial\theta}\left(\sin\theta\frac{\partial}{\partial\theta}\right) + \frac{1}{\sin^2\theta}\frac{\partial^2}{\partial\phi^2}\right] \tag{3.9}$$

whose z-component

$$\hat{L}_z = -i\hbar\frac{\partial}{\partial\phi} \tag{3.10}$$

is also introduced for further convenience. More details about the variable separation and the use of spherical coordinates are found in [3] and [4]. Equation (3.8) makes it clear that *a key role will be played by the angular momentum*, reflecting the central character of $V(r)$.

3.1.2 Stationary states: the wavefunctions

We now aim at solving the equation

$$\hat{H}\psi(r, \theta, \phi) = E\psi(r, \theta, \phi) \tag{3.11}$$

where the Hamiltonian operator is given in equation (3.8) and the stationary wavefunctions are accordingly provided in polar coordinates as indicated in equation (3.7). By inserting such expressions into equation (3.11) we get

$$\frac{1}{r^2}\frac{\partial}{\partial r}\left[r^2\frac{\partial\psi(r,\theta,\phi)}{\partial r}\right] + \frac{1}{r^2\sin\theta}\frac{\partial}{\partial\theta}\left[\sin\theta\frac{\partial\psi(r,\theta,\phi)}{\partial\theta}\right] + \frac{1}{r^2\sin^2\theta}\frac{\partial^2\psi(r,\theta,\phi)}{\partial\phi^2}$$
$$= -\frac{2m_e}{\hbar^2}[E - V(r)]\psi(r,\theta,\phi) \tag{3.12}$$

The strategy for solving such a formidable differential equation fully relies on *the spherical symmetry of the Coulomb potential $V(r)$* which allows one to guess a solution

$$\psi(r, \theta, \phi) = R(r)\Theta(\theta)\Phi(\phi) \tag{3.13}$$

---

[3] Simply, there must be operated the change of variables $x \rightarrow r\sin\theta\cos\phi$, $y \rightarrow r\sin\theta\sin\phi$, and $z \rightarrow r\cos\theta$ in equation (3.3).

where each function only depends upon a sole polar coordinate. While the formal treatment of the stationary Schrödinger equation for the hydrogen atom is outlined in appendix A, here we present just the conceptual scheme of solution, better focussing on the physical significance of the mathematical results.

By inserting the factorization shown in equation (3.13) into equation (3.12) we can obtain, after some non-trivial mathematics, three separate differential equations whose solutions provide the explicit expressions for the $R_{nl}(r)$, $\Theta_{lm_l}(\theta)$, and $\Phi_{m_l}(\phi)$ functions, respectively. Regardless of their analytical form that will be discussed soon, it is important to underline that such functions are characterized by a set of integer numbers $\{n, l, m_l\}$ which are collectively called *quantum numbers*. Mathematics dictates that (see appendix A)

$$
\begin{aligned}
&n = 1, 2, 3, \ldots \\
&l = 0, 1, 2, \ldots, n - 1 \\
&m_l = -l, -l + 1, -l + 2, \ldots, l - 2, l - 1, l
\end{aligned}
\tag{3.14}
$$

It is worth stressing that the discrete nature of the quantum numbers is a natural consequence of the Schrödinger equation. We anticipate that the discreteness of the quantum numbers is linked to the quantization of some corresponding physical quantities: a characteristic that will therefore have to be considered proper for any atomic system (and not postulated ad hoc as it was the case for the electron orbital angular momentum in the atomic Bohr model). It is customary to name n as the *principal quantum number*, l as the *orbital angular momentum quantum number* (or, in short, as the orbital quantum number), and m_l as the *magnetic quantum number*. The reason for such labelling will be clarified shortly. We can accordingly indicate the wavefunctions describing the stationary states as $\psi_{nlm_l}(r, \theta, \phi)$.

3.1.2.1 The angular part
The explicit knowledge of both the Θ- and Φ-functions (see appendix A) allows one to express the *angular part of the stationary wavefunctions* for the hydrogen atom in the form

$$
Y_{lm_l}(\theta, \phi) = \Theta_{lm_l}(\theta)\Phi_{m_l}(\phi)
\tag{3.15}
$$

i.e., in terms of the *spherical harmonic functions* $Y_{lm_l}(\theta, \phi)$ reported in table 3.1; they are associated with two quantum numbers and fulfil the normalization condition

$$
\int_0^{2\pi} d\phi \int_0^{\pi} |Y_{lm_l}(\theta, \phi)|^2 \sin\theta d\theta = 1
\tag{3.16}
$$

It is worth exploiting the physical consequences of what we have found so far: by making use of the spherical harmonic functions, it is cumbersome but straightforward to prove that

$$
\hat{L}^2 \, Y_{lm_l}(\theta, \phi) = l(l + 1)\hbar^2 \, Y_{lm_l}(\theta, \phi)
\tag{3.17}
$$

and

Table 3.1. Some spherical harmonic functions.

l	m_l	$Y_{lm_l}(\theta, \phi)$
0	0	$Y_{00} = \sqrt{1/4\pi}$
1	0	$Y_{10} = \sqrt{3/4\pi}\ \cos\theta$
	± 1	$Y_{1\pm 1} = \mp\sqrt{3/8\pi}\ \sin\theta\ \exp(\pm i\phi)$
2	0	$Y_{20} = \sqrt{5/16\pi}\ (3\cos^2\theta - 1)$
	± 1	$Y_{2\pm 1} = \mp\sqrt{15/8\pi}\ \sin\theta\ \cos\theta\ \exp(\pm i\phi)$
	± 2	$Y_{2\pm 2} = \sqrt{15/32\pi}\ \sin^2\theta\ \exp(\pm 2i\phi)$
3	0	$Y_{30} = \sqrt{7/16\pi}\ (5\cos^3\theta - 3\cos\theta)$
	± 1	$Y_{3\pm 1} = \mp\sqrt{21/64\pi}\ \sin\theta(5\cos^2\theta - 1)\ \exp(\pm i\phi)$
	± 2	$Y_{3\pm 2} = \sqrt{105/32\pi}\ \sin^2\theta\ \cos\theta\ \exp(\pm 2i\phi)$
	± 3	$Y_{3\pm 3} = \mp\sqrt{35/64\pi}\ \sin^3\theta\ \exp(\pm 3i\phi)$

$$\hat{L}_z\ Y_{lm_l}(\theta, \phi) = m_l\hbar\ Y_{lm_l}(\theta, \phi) \tag{3.18}$$

where the operator expressions are given in equations (3.9) and (3.10). Since both $\hat{L}^2$ and $\hat{L}_z$ do not depend on the radial variable, we can more generally write

$$\hat{L}^2\ \psi_{nlm_l}(r, \theta, \phi) = l(l + 1)\hbar^2\ \psi_{nlm_l}(r, \theta, \phi) \tag{3.19}$$

and

$$\hat{L}_z\ \psi_{nlm_l}(r, \theta, \phi) = m_l\hbar\ \psi_{nlm_l}(r, \theta, \phi) \tag{3.20}$$

which is formally consistent with the commuting properties valid for the hydrogen atom[4], namely: $[\hat{H}, \hat{L}^2] = 0 = [\hat{H}, \hat{L}_z]$. It is hard to underestimate the importance of this result: we have demonstrated that *the stationary wavefunctions of the hydrogen atom are simultaneously eigenfunctions of the Hamiltonian operator, of the square orbital angular momentum operator, and of its z-component.* This implies that *in any stationary state the electron has a well-defined value of energy, of square orbital angular momentum, and of z-component of orbital angular momentum.* It is in principle possible to conceive a set of simultaneous measurements of the three observables, returning a trio of well-defined values.

A second major result contained in equations (3.19) and (3.20) is that *the physical observable 'orbital angular momentum' is quantized.* More specifically, quantum mechanics predicts that *the vector module of the electron orbital angular momentum and its component along any arbitrary direction can only assume discrete values.* We stress that we used the label z to indicate such a component, but it obviously corresponds to a totally arbitrary direction in space since, given the spherical

---

[4] More generally, these commuting rules are valid for any quantum problem characterized by a spherical symmetry.

symmetry of the problem, we can arbitrarily orient the frame of reference centred on the nucleus. On the other hand, because of the commuting rules between $\hat{L}_z$ and the other two Cartesian components of the orbital angular momentum operator summarized in equation (2.11), we can state that *in any stationary state the x- and y-components of this observable do not have well-defined values.*

Next, we remark that all possible quantized values of the classical observable L^2 are given by $l(l + 1)\hbar^2$, as appearing in equation (3.17). On one hand, this fully justifies the labelling 'orbital quantum number' for l. On the other hand, it must be observed that for any non-zero value of l there exist $(2l + 1)$ different values of m_l given by equation (3.14).

This state of affairs is summarized by a semi-classical picture according to which in any stationary state of the hydrogen atom with $l \neq 0$ *there is observed a precession of the electron orbital angular momentum vector* **L** *around the (arbitrary) z-direction*, as sketched in figure 3.2 (left): while its projection along z has a well-defined value, both its components on a plane normal to z are ill-defined because of precession. Furthermore, the only possible orientations of **L** with respect to any arbitrary z-direction can be represented as shown in figure 3.2 (right) for the case $l = 1$ and $l = 2$. This picture is useful to get familiar with the most unexpected quantum features of the hydrogen atom, namely: (i) it is impossible to know (i.e., to simultaneously measure) more than one Cartesian component of the electron orbital angular momentum (this feature is represented by the precession); (ii) several possible inclinations of the angular momentum with respect to a given direction are found (this feature is represented by the multiple orientations of **L** each corresponding to a single quantized value of L_z). In brief, we can conclude that *according to quantum mechanics it is impossible to exactly determine the direction of the electron orbital angular momentum.*

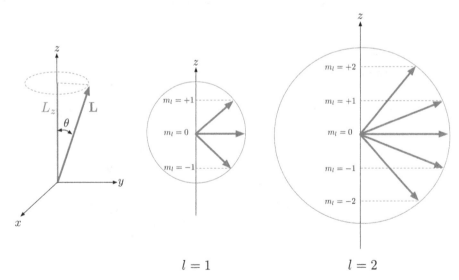

Figure 3.2. Left: pictorial representation of the precession of the orbital angular momentum **L** around the z-axis. Right: the possible orientations of **L** imposed by the quantization rules for the states $l = 1$ and $l = 2$. In both panels the z-axis is arbitrary.

3.1.2.2 The radial part

The radial part of the stationary Schrödinger problem provides the R-functions as well as the energy spectrum for the hydrogen atom (see appendix A). We first devote ourselves to radial functions.

Similarly to harmonic functions, even the radial ones reported in table 3.2 are properly normalized

$$\int_0^{+\infty} |R_{nl}(r)|^2 \, r^2 dr = 1 \tag{3.21}$$

and their shape depends both on l and n. This double dependence determines their mathematical form that can be always put as

$$R_{nl}(r) \sim [\text{polynomial in } r] \times [\text{exponential of } - r] \tag{3.22}$$

The exponential decay ensures that the *wavefunction vanishes at large distances from the atomic nucleus*, as expected: if the electron was likely to be present in that remote region[5], it would be hard to speak of a hydrogen atom, i.e. of a *bound* electron–proton pair. On the other hand, it is a general feature of such radial functions that they have a finite value at the origin for all quantum states characterized by $l = 0$. In other words, we conclude that *the electron accommodated on a stationary state $l = 0$ has a finite probability of being found nearby the nucleus*, at variance with what holds for any other $l \neq 0$ state. This intriguing feature can be explained classically if we make use of a result discussed in appendix A, namely: as for the radial motion, quantum mechanics proves that the electron is subjected to an effective potential

$$W(r) = -\frac{e^2}{4\pi\varepsilon_0}\frac{1}{r} + \frac{l(l+1)\hbar^2}{2m_e}\frac{1}{r^2} \tag{3.23}$$

Table 3.2. Some radial functions of the hydrogen atom expressed in terms of the atomic length unit $a_0 = \varepsilon_0 h^2/\pi m_e e^2$.

n	l	$R_{nl}(r)$
1	0	$2(1/a_0)^{3/2} \exp(-r/a_0)$
2	0	$(1/2a_0)^{3/2} (2 - r/a_0) \exp(-r/2a_0)$
	1	$1/\sqrt{3} \, (1/2a_0)^{3/2}(r/a_0) \exp(-r/2a_0)$
3	0	$2/3 \, (1/3a_0)^{3/2} (3 - 2r/a_0 + 2r^2/9a_0^2) \exp(-r/3a_0)$
	1	$2\sqrt{2}/9 \, (1/3a_0)^{3/2} (2r/a_0 - r^2/3a_0^2) \exp(-r/3a_0)$
	2	$4/(27\sqrt{10}) \, (1/3a_0)^{3/2}(r^2/a_0^2) \exp(-r/3a_0)$

---

[5] We recall the probabilistic interpretation of the wavefunction discussed in section 2.1.2.

where the last term is commonly referred to as the *centrifugal potential*. Let us now imagine an electron approaching the nucleus from a far region: for $l = 0$ there is no centrifugal effect and, therefore, the electron has access to the core nuclear region; on the other hand, if $l \neq 0$ the electron feels a centrifugal barrier pushing it away from the nucleus. The situation is sketched in figure 3.3. While this picture is classical, we can describe the situation in quantum language[6] by saying that the electron wavefunction must vanish outside the classically allowed region. In conclusion, *the larger the value of the orbital quantum number, the smaller is the probability for the electron to approach the nucleus[7]*.

A very effective way to represent the radial behaviour of the electron matter wave is through the *radial distribution function* or *radial probability*

$$P_{nl}(r) = r^2 \mid R_{nl}(r)\mid^2 \tag{3.24}$$

defined as the *probability per unit length that the electron falls at a distance r from the nucleus*. In order to clarify the physical information contained in this function, we consider the stationary state $n = 1$, $l = 0$, $m_l = 0$ and integrate the square modulus of the corresponding total wavefunction (see in equation (3.13) the adopted factorization) over the two polar angles

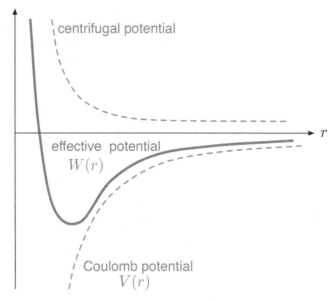

Figure 3.3. Pictorial representation of the Coulomb $V(r) \sim -r^{-1}$ potential, the centrifugal $\sim r^{-2}$ potential, and the effective $W(r)$ potential for a generic state with $l \neq 0$.

---

[6] This way of reasoning will be adopted very often in the future: we will first try to interpret the results according to the canons of classical physics, but then we will translate them into quantum language. Sometimes the opposite procedure will also be followed.

[7] We remark that this conclusion only holds under the assumption of a point-like nucleus.

$$\int_0^{2\pi} d\phi \int_0^{\pi} |\psi_{100}(r, \theta, \phi)|^2 \, r^2 \sin\theta d\theta = r^2 \, | R_{10} |^2 \tag{3.25}$$

which means that the quantity

$$P_{10}(r)dr = | R_{10}(r)|^2 \, r^2 dr \tag{3.26}$$

represents the *probability of finding the electron within the spherical shell centred on the nucleus with radius r and thickness dr*. Similar conclusions hold for any other set of quantum numbers. For any $n > 1$ this specifically means that *the electron matter wave is distributed radially in a way depending on the orbital quantum number*.

The plot of the radial probability distributions is shown in figure 3.4 for $n = 1, 2$ and 3: it provides quite a deep physical insight on the hydrogen atom. By setting the condition

$$\left. \frac{dP_{nl}(r)}{dr} \right|_{r_{max}} = 0 \tag{3.27}$$

we obtain *the distance r_{max} from the nucleus at which the probability of finding the electron is maximum*. It easy to prove that for $n = 1$, $l = 0$, $m_l = 0$ we get $r_{max} = a_0$. In other words, for this stationary state the probability is maximum at a distance corresponding to the radius of the first stable orbit according to the Bohr model discussed in section 1.2.3. While this result in a sense reconciles early atomic physics with rigorous quantum mechanics, we nevertheless underline that in the Bohr model the concept of orbit was still used in a purely classical sense; on the other hand, quantum mechanics, according to its non-deterministic character, only provides a

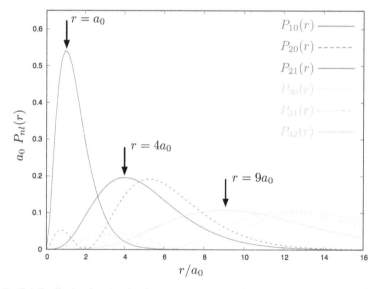

Figure 3.4. Radial distribution function for the states $n = 1, 2, 3$ of the hydrogen atom. For a better graphical representation it has been plotted the dimensionless quantity $a_0 P_{nl}(r)$. Arrows indicate the radii of the first three stable orbits of the Bohr model: they always correspond to maxima of the plotted quantity.

probability of finding the electron at a given distance from the nucleus. Similar considerations are valid for any other quantum state of the hydrogen atom. In figure 3.4 we indicate the distances corresponding to the second and third Bohr stable orbits: interestingly enough, they correspond to maxima of the radial distribution falling at increasingly large l values.

In our full quantum description the concept of orbit is more suitably replaced by the *expectation value* $\langle r \rangle_{nlm_l}$ *of the operator* $\hat{r}$ *associated with the radial electron–proton distance*

$$
\begin{aligned}
\langle r \rangle_{nlm_l} &= \int \psi^*_{nlm_l}(\mathbf{r}) \, \hat{r} \, \psi_{nlm_l}(\mathbf{r}) \, d\mathbf{r} \\
&= \int_0^{+\infty} R^*_{nl}(r) \, \hat{r} \, R_{nl}(r) \, r^2 dr \\
&= \int_0^{+\infty} |R_{nl}(r)|^2 \, r^3 dr \\
&= \left\{ 1 + \frac{1}{2}\left[1 - \frac{l(l+1)}{n^2} \right] \right\} n^2 \, a_0
\end{aligned}
\tag{3.28}
$$

where $d\mathbf{r} = r^2 \sin\theta dr d\theta d\phi$ is the volume element in polar coordinates[8]. Consistently with the above discussion, we argue that equation (3.28) provides *an estimation of the size of the hydrogen atom* that, following the discussion in section 2.4.2, remains constant in time. The atom size is to a good approximation proportional to n^2 since the term $l(l+1)/n^2$ provides comparatively small corrections. It is easy to calculate that $\langle r \rangle_{100} = 3a_0/2 \sim 0.8$ Å, thus quantitatively confirming the order-of-magnitude estimation of the atomic size elaborated in section 1.1.1.

3.1.3 Stationary states: the energy spectrum

The second major result provided by the solution of the radial part of the Schrödinger equation is *the energy spectrum for the hydrogen atom*.

We start from equation (A.17) given in appendix A and consider that it must be $E < 0$ since we are only considering *bound* electron–proton states. By introducing the variable

$$
\xi = \sqrt{\frac{8m_e|E|}{\hbar^2}} \, r
\tag{3.29}
$$

equation (A.17) is reduced to[9]

$$
\frac{d^2u(\xi)}{d\xi^2} + \left[-\frac{l(l+1)}{\xi^2} + \frac{\lambda}{\xi} - \frac{1}{4} \right] u(\xi) = 0
\tag{3.30}
$$

---

[8] In equation (3.28) we have simply outlined the mathematical procedure to calculate $\langle r \rangle_{nlm_l}$. The details of these laborious calculations are omitted for sake of simplicity, but they can be found in any good book of quantum mechanics.
[9] For simplicity we have omitted subscripts.

where

$$\lambda = \frac{e^2}{4\pi\varepsilon_0} \frac{1}{\hbar} \sqrt{\frac{m_e}{2|E|}} \tag{3.31}$$

is a dimensionless quantity. The solutions $u(\xi)$ of equation (3.30) are found in the form of finite non-diverging series only if $\lambda = n$ with $n = 1, 2, 3,...$ [3, 5] a condition that eventually leads to the expression

$$E_n = -\frac{m_e e^4}{8\varepsilon_0^2 h^2} \frac{1}{n^2} = -\mathcal{R}_H hc \frac{1}{n^2} \tag{3.32}$$

defining *the energies of the bound states of the hydrogen atom* in terms of the Rydberg constant $\mathcal{R}_H$ introduced in equation (1.28). A rather striking feature of equation (3.32) is that *energy eigenvalues only depend on the principal quantum number n*, while the full set of three quantum numbers $\{nlm_l\}$ is needed to determine the corresponding eigenfunctions. This means that the hydrogen atom is characterized by a *degenerate energy spectrum*, where[10]

- *the degeneracy in the orbital quantum number l is 'accidental'*: this degeneracy is due to the fact that the Coulomb potential experienced by the electron decays exactly as $\sim 1/r$. Any other central field varying as $\sim 1/r^k$ with $k \neq 1$ generates an energy spectrum E_{nl}, i.e., energy values depending both on n and l, as will be investigated in the case of multi-electron atoms treated within the so-called central-field approximation;
- *the degeneracy in the magnetic quantum number m_l is 'necessary'*: this degeneracy is imposed by the spherical symmetry of the problem. By lowering the degree of symmetry, such a degeneracy is actually removed. This situation is typically found when applying an external electric or magnetic field to the atom, as extensively discussed in the following.

By taking into consideration all the allowed l- and m_l-values corresponding to the same principal quantum number n, we can define the *degree of degeneracy n_d for each quantum stationary state of the hydrogen atom*

$$n_d = \sum_{l=0}^{n-1} (2l + 1) = n^2 \tag{3.33}$$

This expression highlights the fact that *the only non-degenerate state is the ground state $n = 1$*. We are adopting hereafter the notation introduced when developing the Bohr model in section 1.2.3. This is not just a matter of convenience, rather it reflects the fact that quantum mechanics provides the very same electron energies as the heuristic Bohr model. This in turn implies that *equation (3.32) provides full*

---

[10] A more formal treatment of this issue can be found in quantum mechanics textbooks like e.g., [6] or [7].

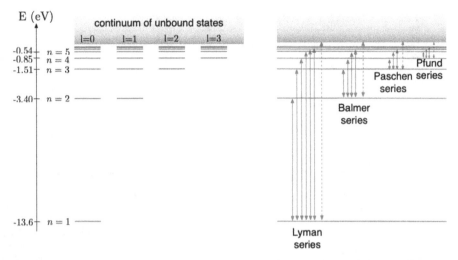

Figure 3.5. Left: energy-level diagram for the hydrogen atom. Right: Map of the radiative transitions observed in atom spectroscopy. The double arrows indicate that each line corresponds to both the absorption and the emission of a photon. The dashed lines mark the convergence limit of each series, corresponding to $n_2 = +\infty$ in equation (1.17).

explanation of the discrete absorption and emission spectra found experimentally and rationalized by the empirical Rydberg formula given in equation (1.17). In figure 3.5 the energy-level diagram for the hydrogen atom (left) and the map of the radiative transitions generating the various spectral series (right) are shown in summary of everything discussed so far.

3.1.4 Classifying the electronic shells

According to a notation widely used in spectroscopy, it is said that all the quantum states with the same n form an *electronic shell* which, therefore, has the property to group all degenerate states with the same energy. Capital letters are used to identify electronic shells according to the correspondence criterium

$$n = 1 \rightarrow \text{shell K} \quad n = 2 \rightarrow \text{shell L} \quad n = 3 \rightarrow \text{shell M} \quad n = 4 \rightarrow \text{shell N}...$$

Similarly, states differing by the orbital quantum number l are labelled by lower letters as

$$l = 1 \rightarrow \text{state } s \quad l = 2 \rightarrow \text{state } p \quad l = 3 \rightarrow \text{state } d \quad l = 4 \rightarrow \text{state } f...$$

They are said to identify the *electronic sub-shells*. Historically, these code letters were introduced as the different spectral lines were experimentally detected and named *sharp, principal, diffuse, fundamental,* and so on.

In summary, we report in table 3.3 the *the classification scheme for the quantum states of the hydrogen atom,* where the ψ-functions are provided in terms of products between radial and spherical harmonic functions, as discussed in the previous sections.

Table 3.3. Classification scheme for the first hydrogen quantum states.

Electronic shell	Set of quantum numbers $\{nlm_l\}$	Spectroscopic notation for the state	Wavefunction labelling
K	$\{100\}$	$1s$	$\psi_{100}(r,\,\theta\,\phi) = \psi_{1s}$
L	$\{200\}$	$2s$	$\psi_{200}(r,\,\theta,\,\phi) = \psi_{2s}$
	$\{210\}$	$2p_0$	$\psi_{210}(r,\,\theta,\,\phi) = \psi_{2p_0}$
	$\{21\pm1\}$	$2p_{\pm1}$	$\psi_{21\pm1}(r,\,\theta,\,\phi) = \psi_{2p_{\pm1}}$
M	$\{300\}$	$3s$	$\psi_{300}(r,\,\theta\,\phi) = \psi_{3s}$
	$\{310\}$	$3p_0$	$\psi_{310}(r,\,\theta\,\phi) = \psi_{3p_0}$
	$\{31\pm1\}$	$3p_{\pm1}$	$\psi_{31\pm1}(r,\,\theta\,\phi) = \psi_{3p_{\pm1}}$
	$\{320\}$	$3d_0$	$\psi_{320}(r,\,\theta\,\phi) = \psi_{3d_0}$
	$\{32\pm1\}$	$3d_{\pm1}$	$\psi_{32\pm1}(r,\,\theta\,\phi) = \psi_{3d_{\pm1}}$
	$\{32\pm2\}$	$3d_{\pm2}$	$\psi_{32\pm2}(r,\,\theta\,\phi) = \psi_{3d_{\pm2}}$

3.1.5 Atomic orbitals

A very appealing feature of the quantum picture is that we can visualize the electron[11] as a *negatively charged cloud* surrounding the nucleus.

Let us consider the generic state $\{nlm_l\}$: since $|\psi_{nlm_l}(\mathbf{r})|^2$ is the probability density of finding the electron at position $\mathbf{r}$, then we can describe it as *a charge distribution with density* $-e|\psi_{nlm_l}(\mathbf{r})|^2$. Once such a charge density is defined, we can look for a surface Σ centred on the nucleus and fulfilling two properties, namely: (i) $|\psi_{nlm_l}(\mathbf{r})|^2 =$ constant for any $\mathbf{r} \in \Sigma$, and (ii) it contains a given fraction ξ of electron charge. It is customarily set $\xi = 0.9$. More specifically, if we consider the state $\psi_{1s}(\mathbf{r})$ then the integral equation

$$\int_0^{R_\Sigma} |\psi_{1s}(\mathbf{r})|^2 \, 4\pi r^2 \, dr = 0.9 \tag{3.34}$$

defines a spherical surface of radius R_Σ containing 90% of the electron charge. The result of the integral appearing in equation (3.34) is $R_\Sigma = 2.6$ Å and we can equivalently say that *the electron is found with 90% probability within such a sphere*. This introduces a very effective way to describe the delocalization of the electron and, in a sense, provides *an intuitive representation of the shape of its quantum state*.

The spherical shape of the surface Σ is a common feature of all *s*-states, although characterized by an increasing value of its radius as n becomes larger and larger. Unfortunately, it is very difficult to calculate and visualize this surface for states with $l > 0$, for which the angular part is provided by complex spherical harmonic functions. To this aim it is better using a different set of wavefunctions which are

---

[11] An electron should be no longer intended as a point-like particle, rather as a matter wave.

built to have a simple and directional dependence on the Cartesian coordinates[12]. By using the set of state vectors $\psi_{nlm_l}(\mathbf{r})$, we can define the following normalized wavefunctions for the shell L

$$\psi_{2p_x} = \frac{1}{\sqrt{2}}\left(\psi_{2p_{+1}} + \psi_{2p_{-1}}\right) = \sqrt{\frac{3}{4\pi}}\, R_{21}(r)\sin\theta\cos\phi$$

$$\psi_{2p_y} = \frac{1}{i\sqrt{2}}(\psi_{2p_{+1}} - \psi_{2p_{-1}}) = \sqrt{\frac{3}{4\pi}}\, R_{21}(r)\sin\theta\sin\phi \tag{3.35}$$

$$\psi_{2p_z} = \psi_{2p_0}$$

that assume their maximum value along the x-, y-, and z-direction, respectively. These new wavefunctions are named *atomic orbitals*. They are still eigenfunctions of both the $\hat{H}$ and $\hat{L}^2$ operator, but *not* of the $\hat{L}_z$ operator as shown explicitly for the ψ_{2p_x} case

$$\hat{L}_z\psi_{2p_x} = -i\hbar\frac{\partial}{\partial\phi}\left[\sqrt{\frac{3}{4\pi}}\, R_{21}(r)\sin\theta\cos\phi\right] = i\hbar\sqrt{\frac{3}{4\pi}}\, R_{21}(r)\sin\theta\sin\phi = i\hbar\psi_{2p_y} \tag{3.36}$$

Atomic $2p$ orbitals are nevertheless very useful in determining the shape of the electronic cloud. By applying to ψ_{2p_x}, ψ_{2p_y}, and ψ_{2p_z} the above definition of the limiting Σ surface we get the result shown in figure 3.6. The same linear combinations reported in equations (3.35) are used for p-states with any arbitrary principal quantum number $n > 1$.

A similar procedure can be followed to generate the so-called d-orbitals. Let us consider the case corresponding to the set $\{n = 3,\ l = 2\}$: we define

$$\psi_{3d_{z^2}} = \psi_{3d_0}$$

$$\psi_{3d_{xz}} = \frac{1}{\sqrt{2}}(\psi_{3d_{+1}} + \psi_{3d_{-1}})$$

$$\psi_{3d_{yz}} = \frac{1}{i\sqrt{2}}(\psi_{3d_{+1}} - \psi_{3d_{-1}})$$

$$\psi_{3d_{xy}} = \frac{1}{i\sqrt{2}}(\psi_{3d_{+2}} - \psi_{3d_{-2}}) \tag{3.37}$$

$$\psi_{3d_{x^2-y^2}} = \frac{1}{\sqrt{2}}(\psi_{3d_{+2}} + \psi_{3d_{-2}})$$

where the subscripts z^2, xz, yz, xy, and $x^2 - y^2$ indicate the behaviour of such orbital functions in Cartesian coordinates. The corresponding limiting surfaces are shown in figure 3.6.

---

[12] In molecular physics this feature will be of great help to describe the directional character of valence chemical bonds.

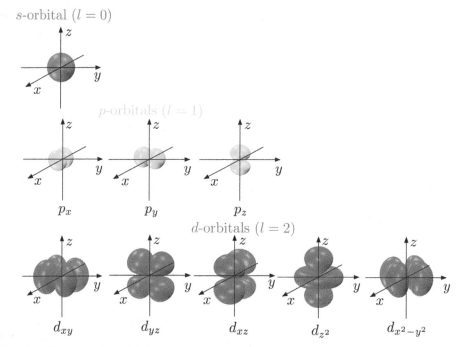

Figure 3.6. Limiting surfaces defining the shape of the atomic orbitals with $l = 0$, 1, and 2 of the hydrogen atom. By courtesy of A Mattoni (CNR-IOM, Cagliari, Italy).

3.2 Hydrogenic atoms

Similarly to the H atom, there exist other atomic systems consisting in just one bound electron–nucleus pair: they are obtained upon ionization of multi-electron atoms. For instance, by removing one electron from helium or two electrons from lithium and so on, we obtain the series of the *hydrogenic atoms* He^+, Li^{2+}, Be^{3+}, ... where the superscripts indicate the degree of ionization.

The main difference with respect to the true H atom consists in the fact that *hydrogenic atoms have a non-unitary nuclear charge* $+Ze$. More specifically, we have $Z = 2$ for He^+, $Z = 3$ for Li^{2+}, $Z = 4$ for Be^{3+}, and so on. Nevertheless, the nucleus can be still considered as a point-like charge and, therefore, it generates a Coulomb field with the very same symmetry characteristics already discussed for the H atom: it is central, although more intense since $Z > 1$. This feature has very important consequences: the factorization of the electron wavefunction into a radial and an angular part, the solution of the Schrödinger equation, the onset of three quantum numbers, and the energy spectrum remain basically unaffected. The only difference is that the atomic number is suitably appearing in some equations. In particular:

- *the angular part of the wavefunction is unaffected*: spherical harmonic functions do not contain any information about the strength of the nuclear field, rather they are only determined by its central character.
- *the radial part of the wavefunction now depends on Z*: with reference to equation (3.22) and table 3.2 written for the true hydrogen atom, this

threefold dependence displays as: (i) the addition of a multiplicative prefactor $Z^{3/2}$; (ii) the replacement of any power r^k appearing in the polynomial term by $(Zr)^k$; and (iii) the replacement of any factor r appearing in the exponential term by (Zr). Another consequence is that the expectation value $\langle r \rangle_{nlm_l}$ given in equation (3.28) becomes inversely proportional to Z, as expected: since this quantity was interpreted as an estimation of the atom size, it is natural to admit that *the larger the nuclear charge, the smaller is such a size*. In other words, the electron being more strongly attracted by larger Z values, its wavefunction tends to shrink in extension. In practice a multiplying factor $1/Z$ must be inserted into equation (3.28).

- *the energy of the quantum states becomes proportional to* Z^2: since the energy eigenvalues of the H atom depend on the square of the nuclear charge, then for hydrogenic atoms equation (3.32) must be simply replaced with

$$E_n = -\frac{m_e Z^2 e^4}{8\varepsilon_0^2 h^2} \frac{1}{n^2} \tag{3.38}$$

Hydrogenic atoms indeed differ from true H even for their nuclear mass[13], which turns out to be increasingly larger than m_p for growing Z values. Therefore, the infinite mass approximation we discussed in section 3.1.1 becomes more and more accurate. Nevertheless, sometimes high-accuracy predictions are needed (or, more simply, we could want to be very rigorous): in these cases, the physics of a hydrogenic atom is better described in terms of an effective negatively charged particle with mass $1/m_{eff} = 1/m_e + 1/M$ (where M is the actual nuclear mass) subjected to a Coulomb field $+Ze/4\pi\varepsilon_0 r$. This reflects in the substitution $m_e \rightarrow m_{eff}$ in equation (3.38) and in the definition of the corrected Bohr radius $\bar{a}_0$

$$\bar{a}_0 = \frac{m_e}{m_{eff}} a_0 \tag{3.39}$$

The new value $\bar{a}_0$ must be used as well in the expressions for the $R_{nl}(r)$ radial functions. This correction to the infinite mass approximation is, for example, used if we need a very accurate prediction of the Rydberg constant $\bar{\mathcal{R}}_H$ for the H atoms which turns out to be

$$\bar{\mathcal{R}}_H = \frac{m_{eff}}{m_e} \mathcal{R}_H = \frac{m_{eff} e^4}{8\varepsilon_0^2 h^3 c} = 109\,681 \text{ cm}^{-1} \tag{3.40}$$

where we have set mass $1/m_{eff} = 1/m_e + 1/m_p$. The result is remarkably close to the experimental measurement reported in equation (1.17).

---

[13] This feature is shared with hydrogen isotopes: deuterium (whose nucleus contains a proton and a neutron) and tritium (whose nucleus contains one proton and two neutrons).

3.3 Magnetic moments and interactions

As anticipated, an electron moving around a proton along a closed orbit corresponds to a current, as shown in figure 3.7. While this is a classical picture, it is worthy of further investigation since it is a promising approach to introduce magnetic interactions into a more refined description of the hydrogen atom.

Let us consider a charge $-e$ moving along a closed, stable, and circular orbit with radius r. If the electron moves with a constant (linear) velocity v, then the modulus of the corresponding orbital angular momentum is $L = m_e vr$ and the associated current is $i = ev/2\pi r$. Ampère's principle of equivalence allows associating a *magnetic dipole moment* with the current

$$M_L = i\pi r^2 = \frac{1}{2}evr = \frac{e}{2m_e}m_e vr = \frac{e}{2m_e}\,L \qquad (3.41)$$

where the subscript L indicates that such a magnetic moment is associated with the electron orbital motion. For historical reasons this equation is usually set in a different form

$$M_L = \frac{e}{2m_e}\,L = \frac{e\hbar}{2m_e}\,g_L\,\frac{1}{\hbar}L \qquad (3.42)$$

where the constant $g_L = 1$, admittedly not necessary at this stage, is named *orbital g-factor* and will find full explanation later on. The quantity

$$\mu_B = \frac{e\hbar}{2m_e} = 9.274\,010 \times 10^{-24}\ \text{J T}^{-1} \qquad (3.43)$$

is called *Bohr magneton*: since the ratio $L/\hbar$ is dimensionless, it plays the role of the 'unit of magnetic moment' in atomic physics (it is assumed to measure the magnetic field in Tesla units, symbol T).

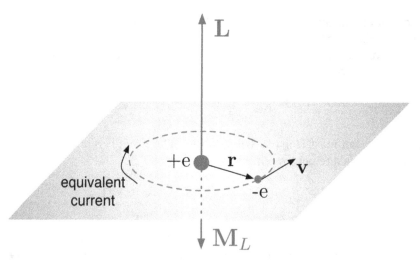

Figure 3.7. Classical picture introducing the concept of electron orbital magnetic moment $\mathbf{M}_L$.

While the vectors $\mathbf{M}_L$ and $\mathbf{L}$ have the same direction (i.e., they are normal to the plane containing the orbit), they have opposite orientation, as shown in figure 3.7 and, therefore, we can write

$$\mathbf{M}_L = -g_L \, \frac{\mu_B}{\hbar} \, \mathbf{L} \tag{3.44}$$

We now observe that the *gyromagnetic ratio* M_L/L does not depend on the orbital radius, nor on the orbital frequency, nor on the actual shape of the orbit itself (in fact, its definition does not require the notion of 'electron orbit'). We conclude that the form given in equation (3.44) for the magnetic moment does not depend on the specific model we used to derived it. In other words, *a full quantum treatment* (where no details about the orbit are in fact used just because the very concept of orbit is ill-defined) *would lead to the same result*. In order to switch from the classical picture to the quantum one, we must simply adopt the well-known quantization rules for the angular momentum. Accordingly, we get

$$M_L = \frac{g_L \mu_B}{\hbar} \, \sqrt{l(l+1)} \, \hbar = g_L \, \mu_B \, \sqrt{l(l+1)}$$
$$M_{L,z} = -\frac{g_L \mu_B}{\hbar} \, m_l \hbar = -g_L \, \mu_B \, m_l \tag{3.45}$$

for the modulus and the z-component of the electron orbital magnetic moment, respectively. This is a very important result since it states that *a hydrogen atom is equivalent to a quantized magnetic dipole*. An external magnetic field is therefore expected to interact with the atom: we are now interested in understanding the new physics emerging from such coupling.

3.3.1 The action of a uniform magnetic field: the Zeeman effect

Let us suppose that a uniform and constant magnetic field $\mathbf{B}$ is applied to an hydrogen atom. The field is assumed parallel to the z-axis, i.e., $\mathbf{B} = (0, 0, B_z)$, forming an angle θ with the atomic magnetic moment. While no translational motion is produced, the action is twofold, namely (i) the atom undergoes a torque

$$\mathbf{T} = \mathbf{M}_L \times \mathbf{B} = -\frac{g_L \mu_B}{\hbar} \, \mathbf{L} \times \mathbf{B} \tag{3.46}$$

and (ii) an additional magnetic potential energy

$$E_{\text{mag}} = -\mathbf{M}_L \cdot \mathbf{B} \tag{3.47}$$

is added to the orbiting electron.

The torque makes the angular momentum $\mathbf{L}$ precess around the direction of the magnetic field according to the fundamental equation of rotational dynamics

$$\frac{d\mathbf{L}}{dt} = \mathbf{M}_L \times \mathbf{B} \tag{3.48}$$

If ω is the precession angular frequency, then the angular momentum is varied by the amount $dL = L \sin \theta \ \omega dt$ over an infinitesimal time span dt, as sketched in figure 3.8. Therefore, it holds that

$$L \sin \theta \ \omega = \frac{dL}{dt} = M_L B \sin \theta \tag{3.49}$$

which leads to

$$\omega_{\text{Larmor}} = \frac{g_L \mu_{\text{B}}}{\hbar} B \tag{3.50}$$

known as the *Larmor precession frequency*.

More interestingly, the action of the magnetic field lowers the symmetry of the atomic problem from spherical to axial and, as anticipated, this *removes the degeneracy in the magnetic quantum number*. For the conditions defined above, the quantum expectation value of the magnetic potential energy is[14]

$$\langle E_{\text{mag}} \rangle = g_L \ \mu_{\text{B}} \ B_z \ m_l \tag{3.51}$$

which implies that the $(2l + 1)$ levels that are degenerate for $\mathbf{B} = 0$ are now split into equidistant levels separated by an energy gap $\Delta E_{\text{mag}} = \mu_{\text{B}} B_z$. This is called *normal Zeeman effect*[15] and the related effect on the energy spectrum is referred to as *Zeeman splitting*. The situation is sketched in figure 3.9 in the case of $l = 0$, 1, and 2. Of course, no Zeeman effect is expected for the s-state. It is interesting to remark that

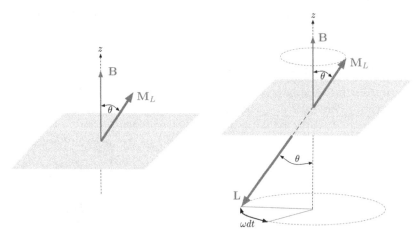

Figure 3.8. Schematic representation of the Larmor precession of the angular **L** and magnetic $\mathbf{M}_L$ moments under the action of a uniform and constant magnetic field **B**. ω is the Larmor precession frequency and dt is an infinitesimal time interval.

---

[14] This expectation value is calculated on the hydrogen wavefunctions described in section 3.1.2.
[15] The need of such labelling will be clarified in the following, when the so-called 'anomalous' Zeeman effect will be described in connection to the electron spin.

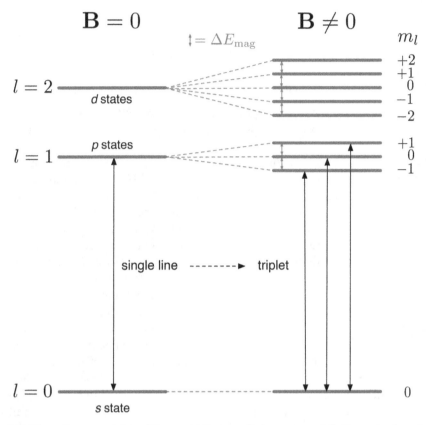

Figure 3.9. Schematic representation of the normal Zeeman effect on s-, p-, and d-states under the action of a uniform and constant magnetic field $\mathbf{B} = (0, 0, B_z)$. The quantity $\Delta E_{\mathrm{mag}} = \mu_B B_z$ is called Zeeman splitting. It is shown that the single emission/absorption line observed for the $s \leftrightarrow p$ transition with no magnetic field is resolved into a triplet.

the Zeeman splitting does not depend on the n and l quantum numbers: this implies that all states undergo the very same separation under the action of a magnetic field.

The splitting has important consequences in spectroscopy: for instance, the single emission/absorption line observed in the absence of any magnetic field for the transition occurring between an s-state and a p-state is now replaced by three Zeeman components which are usually referred to as a *triplet*. Similar features are found for other transition (single lines are possibly opened in multiplets as, e.g., in the case of the transition between a p-state and a d-state). All these Zeeman-like features are summarized in figure 3.9.

In conclusion, we can attempt a first order-of-magnitude evaluation of the magnetic interactions in atomic systems. If we assume an electron magnetic moment of $\sim 1 \ \mu_B$ and a magnetic field of ~ 1 T (which is in fact a somewhat intense field, if one recalls that the Earth's magnetic field ranges in between 20 and 60 μT), we obtain $\Delta E_{\mathrm{mag}} \sim 10^{-4}$ eV, indeed very much smaller than the energy separation between consecutive hydrogen-like quantum levels obtained by only considering

Coulomb interactions, which typically fall in the ~0.1–10 eV range. This justifies our choice to neglect in the first instance any magnetic features when setting up the quantum mechanical problem for the hydrogen atom.

3.3.2 The action of a non-uniform magnetic field: the electron spin

Let us now turn to the more intriguing case of a *non-uniform* (but still constant in time) magnetic field acting, say, along the z-direction. Let $\mathbf{B} = (0, 0, B_z(z))$ be its vector representation.

When a beam of atoms collimated perpendicularly to $\mathbf{B}$ enters the region where the field gradient occurs, it is subjected to a force $F_z = M_{L,z}\, dB_z/dz$ and, therefore, *it is transversally deflected*. The actual deflection depends on the z-component of the quantized magnetic moment of its atoms. This situation is realized by the apparatus schematically shown in figure 3.10, originally set up by O Stern and W Gerlach in 1924. Historically the Stern–Gerlach experiment was performed with silver atoms, an atomic system we are not yet prepared to deal with. However, there is nothing that conceptually prevents an experiment of this type from being carried out with a beam of hydrogen atoms[16]. More specifically, we can imagine collimating a beam of H atoms prepared in a quantum s-state, i.e., in a state with $l = 0 = m_l$. Therefore, although there is a magnetic field gradient, the resulting deflecting force would be zero. In other words, the quantum prediction is that such a beam would emerge from the Stern–Gerlach apparatus without having undergone any deflection. If by chance we did some experimental error in preparing the atomic beam (i.e., the actual quantum states was not s and, therefore, $m_l \neq 0$), then quantum mechanics predicts that the incoming beam would split into *an odd number of emerging beams*, reflecting the $(2l + 1)$ quantized z-components of the electron magnetic moment. On the other

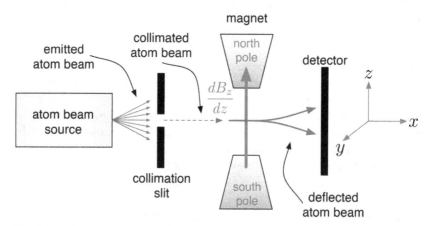

Figure 3.10. Schematic representation of the Stern–Gerlach apparatus for measuring the deflection of an atomic beam by a non-uniform magnetic field normal to the beam propagation direction.

---

[16] The reason why working with H atoms is unpractical is that they are very light objects and, therefore, any possible deviation of their trajectory would be so small as to be barely detectable.

hand, if classical physics was at work here, *the emerging beam would appear as a continuum*, reflecting the arbitrary direction of such a moment[17].

The result is completely at odd with expectations: in fact, they are observed as *just two beams, symmetrically deflected with respect to the incoming direction.* Furthermore, it is observed that the separation corresponds to a magnetic moment of ~1 μ_B. Clearly this experimental finding is not consistent with classical physics (predicting a continuum of deflected beams) nor with quantum mechanics as developed so far (predicting either zero deflection for *s*-states or an odd $(2l + 1)$ number of deflected beams for *p*-, *d*-, ... states)[18].

These intriguing results were explained in 1926 by S A Goudsmit and G E Uhlenbeck by assuming that *the electron is characterized by an additional intrinsic angular momentum*, named *spin* and labelled **S**, such that: (i) it obeys the ordinary quantization rules holding in quantum mechanics for angular momenta; (ii) its square modulus and *z*-component are provided by the following equations

$$S^2 = s(s + 1)\, \hbar^2$$
$$S_z = m_s\, \hbar$$

(3.52)

where the possible values of m_s differ by one unit and are contained in the interval $[-s, +s]$; and (iii) an *intrinsic magnetic moment* $\mathbf{M}_S$ is associated to the spin degree of freedom according to

$$\mathbf{M}_S = -g_S\, \frac{\mu_B}{\hbar}\mathbf{S}$$

(3.53)

where g_S is named *spin g-factor*. Since the Stern–Gerlach experiment provides evidence for just two beams deflected by the non-uniform magnetic field, then it must hold

$$m_s = \pm\frac{1}{2} \quad \rightarrow \quad s = \frac{1}{2}$$

(3.54)

Let us now return to the experiment: thanks to the Goudsmit–Uhlenbeck hypothesis, we can admit that any atom was correctly prepared in an *s*-state, still having a non-zero magnetic moment provided by equation (3.53). Therefore, we can calculate the intensity of the deflecting force as

$$F_z = M_{S,z}\, \frac{dB_z}{dz} = g_S\, \frac{\mu_B}{\hbar}\, m_s\hbar\frac{dB_z}{dz}$$

(3.55)

where both F_z and dB_z/dz are known since they are, respectively, measured and imposed. Since, as reported above, the separation between the two deflected beams is

---

[17] Historically the Stern–Gerlach experiment was designed as a direct test of the space quantization of the angular momentum.

[18] We further observe that this result is not explained by the existence of a nuclear magnetic moment, as found in experimental nuclear physics. For a nucleus of mass M the corresponding magnetic moment is of the order of $e\hbar/2M \ll 1\mu_B$ since $M \gg m_e$.

~1 μ_B, it is eventually found that $m_s g_S = \pm 1$, from which we predict the value of the spin g-factor to be $g_S = 2$.

The true origin[19] of the new intrinsic degree of freedom named spin can only be accounted for by a full relativistic quantum theory [8]. Historically, the spin was tentatively associated with the rotation of the electron around its axis: since such a rotation can only occur clockwise or counter-clockwise, this simple model deceptively suggested why there are only two possible spin values; however, the quantitative exploitation of this model was badly unsuccessful, as reported in section 3.3.3. There is no other conclusion: *at the non-relativistic level, the most clean approach to spin is simply to admit its existence by experimental evidence.* Interestingly enough, while the non-relativistic quantum formalism does not predict the existence of the spin, it is fully compatible with it. We simply need to extended the set of electron degrees of freedom so as to include the spin one: the hydrogenic electron wavefunction will accordingly be written as a *spin-orbital*

$$\psi_{nlm_l s}(\mathbf{r}; s) = \psi_{nlm_l}(\mathbf{r}) \, \chi_{sm_s} \tag{3.56}$$

where the χ-function represents the eigenfunction of the spin angular momentum operators

$$\begin{aligned}
\hat{S}^2 \, \chi_{sm_s} &= s(s+1)\hbar^2 \, \chi_{sm_s} \\
\hat{S}_z \, \chi_{sm_s} &= m_s \hbar \chi_{sm_s}
\end{aligned} \tag{3.57}$$

and the *spin quantum numbers* s and m_s are provided in equation (3.54). In conclusion, we can state that, by including the spin in our physical picture, *each hydrogenic quantum state is $2n^2$-fold degenerate.*

The actual existence of the spin and the validity of the related formal rules are confirmed by investigating the energy spectrum of the hydrogen atom under the action of an external magnetic field. Let us consider the hydrogen ground state $n = 1$, $l = 0$, and $m_l = 0$. By applying a uniform and constant magnetic field $\mathbf{B} = (0, 0, B_z)$ it is observed that such a level E_{100} is symmetrically split into two sub-levels: the splitting ΔE is due to coupling of the field to spin moments with opposite directions. By calculating the quantum expectation value

$$\langle -\mathbf{M}_S \cdot \mathbf{B} \rangle = -M_{S,z}B_z = g_S \mu_B m_s B_z = \pm \frac{1}{2} g_S \mu_B B_z \tag{3.58}$$

where the $\pm$ sign indicates that the electron spin could be either parallel or anti-parallel to the field, we predict

$$\Delta E = g_S \mu_B B_z \tag{3.59}$$

as schematically reported in figure 3.11. We remark that the symmetrical splitting is the signature of the twofold value of the spin $s = \pm 1/2$: the new energy spectrum is,

---

[19] The emerging of the spin is a natural consequence of the request of relativistic invariance of the Schrödinger equation describing an electron embedded in an electromagnetic radiation field.

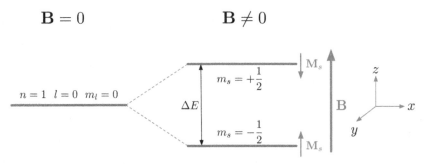

Figure 3.11. Fine structure of the hydrogen ground state ($n = 1$, $l = 0$, and $m_l = 0$) under the action of a magnetic field **B**. The two possible alignments of the spin magnetic moment $\mathbf{M}_S$ with respect to the magnetic field **B** are shown on the right.

therefore, a robust confirmation of the Goudsmit–Uhlenbeck hypothesis. The most accurate measurements actually provide the best estimate for the spin g-factor to be $g_S = 2.002\ 32$.

The experimental validation of equation (3.59) is found similarly to the case of the Zeeman effect. Let us focus on the example illustrated in figure 3.11: the interaction of the spin with an external magnetic field resolves the spin degeneracy of the $(n, l, m_l) = (1, 0, 0)$ level which opens in two sub-levels corresponding to $m_s = \pm 1/2$. This implies that any absorption spectral lines with initial state $(n, l, m_l) = (1, 0, 0)$ is now split into a doublet. This can be addressed by accurate spectroscopic measurements which, detecting the frequency separation between the two lines of the doublet, can provide a direct estimation of ΔE.

3.3.3 The spin is not a classical degree of freedom

As previously anticipated, at the dawn of the quantum era (still largely inspired to classical physics) it was natural to speculate that the spin degree of freedom is associated with some intrinsic rotational motion of the electron. To fully exploit this picture, we describe the electron as a charged sphere of radius a and mass m_e rotating about an axis passing through its centre with angular velocity ω.

Classical mechanics [9] provides the electron angular momentum S of rotation

$$S = I\omega \qquad (3.60)$$

where

$$I = \frac{2}{5}m_e a^2 \qquad (3.61)$$

is its moment of inertia. Let us now assume that the charge is uniformly distributed with density σ just at the surface of the electron. If we resolve its spherical surface into infinitesimal circular crowns of internal radius $a \sin \theta$ and thickness $a\,d\theta$ (where $0 \leqslant \theta \leqslant \pi$ is the angle formed between the axis of rotation and the radius identifying the position of the crown), we can easily calculate the corresponding magnetic moment

$$dM_S = A di = \pi(a \sin \theta)^2 \ di \qquad (3.62)$$

where

$$di = \frac{dq}{dt} = \frac{\sigma(2\pi a \sin \theta)(a d\theta)}{2\pi/\omega} = \sigma \omega a^2 \sin \theta d\theta \qquad (3.63)$$

is the infinitesimal current flowing in the crown. This results into a total spin magnetic moment

$$M_S = \int dM_S = \pi a^4 \sigma \omega \int_0^\pi \sin^3 \theta d\theta \qquad (3.64)$$

which, by solving the integral[20], leads to

$$M_S = \frac{4}{3}\pi a^4 \sigma \omega = \frac{(4\pi\sigma a^2)(a^2\omega)}{3} = \frac{q(a^2\omega)}{3} \qquad (3.65)$$

where q is the total charge distributed on the surface of the electron sphere. Let us now calculate the *classical spin gyromagnetic ratio* by using equations (3.60) and (3.65)

$$\gamma_{\text{classical}} = \frac{M_S}{S} = \frac{5}{6}\frac{q}{m_e} = 1.46 \times 10^{11} \frac{C}{kg} \qquad (3.66)$$

where we have of course set $q = e$. On the other hand, by means of equation (3.53) the quantum mechanical prediction for this ratio is

$$\gamma_{\text{quantum}} = \frac{g_S \ \mu_B}{\hbar} = 1.75 \times 10^{11} \frac{C}{kg} \qquad (3.67)$$

showing a strong disagreement (as large as 20%) with the classical prediction.

We might ask whether the disagreement between classical and quantum theory is due to the assumption that the electron charge is distributed only on its surface. We therefore change this picture by now assuming that such a charge is distributed within the entire volume of the sphere; in order to make things simple, we further guess a uniform charge density $\rho = 3q/4\pi a^3$. In this case, it is convenient to resolve the electron volume into concentric spherical shells with increasing radius r, thickness dr, and all rotating with the same angular frequency ω. The current produced by the rotating charge dq contained in each shell generates an infinitesimal magnetic moment

$$dM_S = \frac{1}{3}\omega(dq)r^2 \qquad (3.68)$$

---

[20] This integral is calculated as $\int_0^\pi \sin^3 \theta d\theta = \int_0^\pi \sin^2 \theta \sin \theta d\theta = -\int_1^{-1}(1 - x^2)dx = 4/3$ where we have set $\sin \theta d\theta = -d(\cos \theta)$ and $\sin^2 \theta = (1 - \cos^2 \theta)$.

with

$$dq = 4\pi\rho r^2 dr \tag{3.69}$$

which allows one to calculate

$$M_S = \int dM_S = \frac{4\pi}{3}\omega\rho \int_0^a r^4 dr = \frac{q\omega a^2}{5} \tag{3.70}$$

as the total spin magnetic moment of the electron. Once again, in this equation it is understood $q = e$. Eventually we get $\gamma_{classical} = 8.78 \times 10^{10}$ C kg^{-1}, even worse in disagreement with the quantum result than in the previous calculation.

The key conclusion is: for whatever distribution of the electron charge, we cannot explain the experimental evidences for the spin gyromagnetic ratio by classical physics. Therefore, we eventually conclude that *the spin cannot be identified with a classical rotational degree of freedom*.

3.4 Spin–orbit interaction

Similarly to the orbital magnetic moment (see section 3.3.1), even the spin one interacts with a magnetic field which, in principle, could be external or internal, i.e. *generated by the relative electron–nucleus orbital motion*. While the first situation will be addressed in the following, here we focus on the case of *magnetic interactions occurring between the electron spin and the orbital current* in a hydrogenic atom with nuclear charge $+Ze$.

The situation is conceptualized in figure 3.12 with respect to a frame of reference centred on the nucleus (left) or on the electron (right). By opting for the latter choice, we can state[21] that the electron is subjected to action of a magnetic field

$$\mathbf{B} = \frac{\mu_0}{4\pi} \frac{(-Ze\mathbf{v}) \times \mathbf{r}}{r^3} \tag{3.71}$$

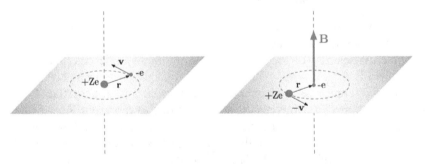

Figure 3.12. The relative electron–nucleus orbital motion represented in a frame of reference centred on the nucleus (left) or on the electron (right). Here **B** is the magnetic field experienced by the electron because of the relative motion of the nucleus.

---

[21] We make use of the Biot–Savart law [2]: an infinitesimal trajectory segment $d\mathbf{l}$ crossed by a steady current i generates an infinitesimal magnetic field $d\mathbf{B} = (\mu_0/4\pi)(id\mathbf{l} \times \mathbf{r})/r^3$, where μ_0 is the vacuum magnetic permeability.

generated by the nuclear motion with the orbital speed $-v$. The electron also undergoes the action of the nuclear electric field

$$\mathbf{E} = \frac{1}{4\pi\varepsilon_0} \frac{Ze}{r^3} \mathbf{r} \tag{3.72}$$

We can therefore write

$$\mathbf{B} = \frac{\mu_0}{4\pi} \frac{(-Zev) \times \mathbf{r}}{r^3} = -\mu_0\varepsilon_0 \, v \times \left(\frac{1}{4\pi\varepsilon_0} \frac{Ze}{r^3} \mathbf{r} \right)$$
$$= -\mu_0\varepsilon_0 \, v \times \mathbf{E} = -\frac{1}{c^2} \, v \times \mathbf{E} \tag{3.73}$$

where $c = 1/\sqrt{\mu_0\varepsilon_0}$ is the speed of light. This expression can be usefully manipulated by observing that if we indicate with $\mathbf{F}$ and $V(r)$, respectively, the nuclear Coulomb force on the electron and the corresponding electrostatic potential, then

$$\mathbf{E} = -\frac{\mathbf{F}}{e} = \frac{1}{e} \frac{dV(r)}{dr} \frac{\mathbf{r}}{r} \tag{3.74}$$

and therefore

$$\mathbf{B} = -\frac{1}{c^2} \, v \times \mathbf{E} = -\frac{1}{c^2} \, v \times \left[\frac{1}{e} \frac{dV(r)}{dr} \frac{\mathbf{r}}{r} \right]$$
$$= -\frac{1}{ec^2} \frac{dV(r)}{dr} \frac{1}{r} \, v \times \mathbf{r}$$
$$= -\frac{1}{em_ec^2} \frac{dV(r)}{dr} \frac{1}{r} \, m_e v \times \mathbf{r}$$
$$= \frac{1}{em_ec^2} \frac{dV(r)}{dr} \frac{1}{r} \, \mathbf{L} \tag{3.75}$$

This result clearly indicates that *the atomic internal magnetic field is generated by the orbital motion* described by the angular momentum $\mathbf{L}$.

The potential energy of the spin magnetic moment under the action of such a magnetic field is $E_{mag} = -\mathbf{M}_S \cdot \mathbf{B}$. We remark that this result has been obtained in the frame of reference centred on the electron, which is not inertial. For sake of consistency with any previous result, it is preferable to represent E_{mag} in the inertial frame of reference centred on the nucleus: this is tantamount to performing a Lorentz transformation between the two situations sketched in figure 3.12. This tedious (but not complicated) calculation[22], leads to a physically very clean result: an observer sitting in the fixed frame of reference centred on the nucleus sees any other frame of reference centred on the rotating electron to precess relative to his/her own

---

[22] This calculation is for instance reported in [4] where the Thomas precession is obtained by a relativistic velocity transformation or in [10] where to get the same result an argument based on relativistic time dilatation is used.

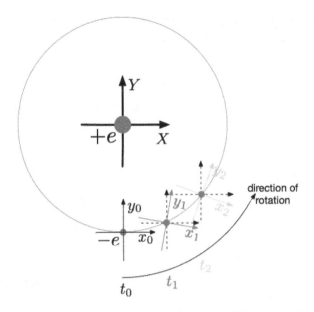

Figure 3.13. Black thick lines: frame fixed on the nucleus $+e$ (top view with respect to the planar circular orbit shown in magenta color). Thin color lines: frames centred on the moving electron $-e$: they appear rotated because of Thomas precession, when observed from the $\{X,\ Y\}$ frame at different $t_0 < t_1 < t_2$ times.

set of axes. The situation is graphically represented in figure 3.13. It is proved [4, 10] that the corresponding *Thomas precession frequency* is

$$\omega_{\text{Thomas}} = -\frac{1}{2c^2}v \times \mathbf{a} \tag{3.76}$$

where $\mathbf{a} = -e\mathbf{E}/m_e$ is the electron acceleration toward the nucleus. Consequently, in the fixed frame of reference centred on the nucleus *the spin magnetic moment is seen to precess with a total angular frequency*

$$\omega = \omega_{\text{Larmor}} + \omega_{\text{Thomas}} \tag{3.77}$$

where the first term is calculated as in equation (3.50) by replacing g_L with the spin g-factor and by using equation (3.73) for the internal magnetic field. An explicit calculation easily leads to

$$\omega = \underbrace{-\frac{e}{m_e c^2}v \times \mathbf{E}}_{\text{Larmor precession}} + \underbrace{\frac{e}{2m_e c^2}v \times \mathbf{E}}_{\text{Thomas precession}} = -\frac{e}{2m_e c^2}v \times \mathbf{E} \tag{3.78}$$

which proves that in passing from the frame of reference centred on the electron (where the electron only undergoes Larmor precession) to the fixed frame of reference centred on the nucleus (where the electron also experiences relativistic effects) a simple corrective 1/2 factor appears. The same conclusion holds for the orientational energy E_{mag} of the spin magnetic moment since its magnitude is proportional to the precession frequency. Accordingly, *the spin magnetic potential energy calculated in the fixed frame of reference centred on the nucleus is*

$$E_{\mathrm{mag}} = -\frac{1}{2}\,\mathbf{M}_S \cdot \mathbf{B} \tag{3.79}$$

where the 1/2 factor is customarily referred to as the *Thomas precession correction term*. Eventually, by inserting into this equation the expression for the magnetic field provided in equation (3.75) we get

$$
\begin{aligned}
E_{\mathrm{mag}} &= -\frac{1}{2}\,\mathbf{M}_S \cdot \mathbf{B} \\
&= -\frac{1}{2}\left(-\frac{g_S \mu_{\mathrm{B}}}{\hbar}\right)\frac{1}{em_e c^2}\frac{dV(r)}{dr}\frac{1}{r}\,\mathbf{S} \cdot \mathbf{L} \\
&= \frac{1}{2m_e^2 c^2}\frac{1}{r}\frac{dV(r)}{dr}\,\mathbf{S} \cdot \mathbf{L}
\end{aligned}
\tag{3.80}
$$

where we have made use of the fact that $g_S = 2$ and $\mu_{\mathrm{B}} = e\hbar/2m_e$. By setting

$$\xi(r) = \frac{1}{2m_e^2 c^2}\frac{1}{r}\frac{dV(r)}{dr} \tag{3.81}$$

we can write the expression of the *spin–orbit interaction energy in hydrogenic atoms* as

$$E_{\mathrm{so}} = \xi(r)\,\mathbf{S} \cdot \mathbf{L} \tag{3.82}$$

where we replaced the subscript 'mag' appearing in equation (3.80) by the subscript 'so' to highlight the internal origin of this energy contribution, which is only due to the interaction between the spin magnetic moment and the magnetic field generated by relative electron–nucleus motion.

It is interesting to attempt an order-of-magnitude estimation of the spin–orbit interaction energy. From equation (3.82) we understand that s-states are not affected by spin–orbit coupling since $\mathbf{L} = 0$. Therefore, we can address the specific case of the state $n = 2$ with $l = 1$ in the hydrogen atom. In this case $V(r) = -e^2/4\pi\varepsilon_0 r$ so that $\xi(r) = \bar{\xi} = e^2/8\pi m_e^2 c^2 \varepsilon_0 (4a_0)^3$ where, to a good approximation, we have set $r \sim 4a_0$, i.e., the distance at which the probability of finding the electron is maximum, as shown in figure 3.4. As for the $\mathbf{S} \cdot \mathbf{L}$ term we take $S = \sqrt{3}\,\hbar/2$ and $L = \sqrt{2}\,\hbar$ resulting in $\mathbf{S} \cdot \mathbf{L} \sim \hbar^2$. In summary, $E_{\mathrm{so}} \sim \pm\bar{\xi}\hbar^2$ which generates a spin–orbit splitting $\Delta E_{\mathrm{so}} \sim 2\bar{\xi}\hbar^2 \sim 10^{-4}$ eV. Similarly to the case of the Zeeman splitting, even the internal magnetic interactions provide energy separations which are comparatively much smaller than the ones associated with electrostatic coupling. This order-of-magnitude estimate suggests that *we can treat spin–orbit effects as perturbations to the energy spectrum obtained by solving equation (3.12)*. More specifically, we guess that the spin–orbit interaction does not mix the eigenstates of an hydrogenic atom and, therefore, we can still use in the first instance its wavefunctions (see sections 3.1.2 and 3.2) to describe them: therefore, *spin–orbit effects will be treated according to the quantum perturbation theory* described in section 2.7.

In order to quantitatively estimate the spin–orbit coupling, we need to calculate separately the quantum expectation value of $\xi(r)$ term and the $\mathbf{S} \cdot \mathbf{L}$ product appearing in equation (3.82). In the first case, we can proceed as reported in appendix B according to which the expectation value for the spin–orbit coupling constant for hydrogenic atoms is

$$\langle \xi(r) \rangle = \frac{1}{2m_e^2 c^2} \left(\frac{e^2}{4\pi\varepsilon_0} \right) \frac{Z^4}{a_0^3} \frac{1}{n^3 \, l\left(l + \frac{1}{2} \right)(l + 1)} \tag{3.83}$$

This is a truly remarkable achievement since it proves that spin–orbit effects are (i) *more sizeable for heavy atoms with large Z values* and (ii) *depend on both the principal and angular quantum numbers*[23].

The evaluation of the $\mathbf{S} \cdot \mathbf{L}$ product proceeds in a rather different way: we are anticipating some feature of the semi-classical 'vector model' of the atom that will be extensively used in section 5.4 for treating magnetic interactions in multi-electron atoms. Without spin–orbit coupling the $\mathbf{L}$ and $\mathbf{S}$ angular momenta would be independent or, equivalently, the observables L^2, S^2, L_z, and S_z would simultaneously have well-defined values. On the other hand, through the spin–orbit interaction the two momenta are coupled in such a way that they both precess around the direction of their vector sum $\mathbf{J} = \mathbf{L} + \mathbf{S}$, as shown in figure 3.14.

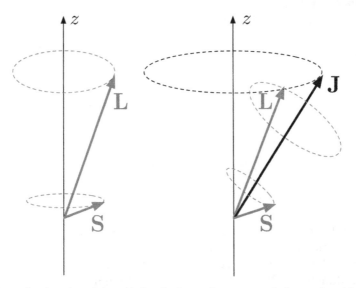

Figure 3.14. Precessional motion of the orbital and spin angular momenta. Left: no spin–orbit interaction is taken into account. Right: spin–orbit coupling is duly taken into account.

---

[23] It should be noted that this result is valid under the assumption of infinite nuclear mass and, therefore, it can be corrected to include finite-mass effects by using equation (3.39).

The origin of such precessional motion is easily understood in terms of torque action, as already discussed in the case of the Zeeman effect (see figure 3.8): the internal magnetic field, which depends on **L**, produces a torque on $\mathbf{M}_S$, which depends on **S**. We can immediately switch to the quantum formalism by assuming that the new *total angular momentum* **J** does fulfil the quantization rules holding for any angular momentum vector, namely

$$J^2 = j(j + 1)\ \hbar^2$$
$$J_z = m_j\ \hbar \tag{3.84}$$

where we have introduced two new quantum numbers j and m_j that, in presence of spin–orbit coupling, must be referred to as the real 'good quantum numbers' for our problem. The quantum theory of angular momenta is naturally extended to **J** and, therefore, we can write its eigenvalue operator equations

$$\hat{J}^2 \psi_{nlm_l} = j(j + 1)\hbar^2\ \psi_{nlm_l}$$
$$\hat{J}_z\ \psi_{nlm_l} = m_j\hbar\psi_{nlm_l} \tag{3.85}$$

where of course $m_j = -j, -j + 1, -j + 2, \ldots, j - 2, j - 1, j$. Since $J_z = L_z + S_z = m_l\hbar + 1/2\hbar$, the maximum possible value of m_j is $(l + 1/2)$, l being the maximum value for m_l. Interestingly enough, this $(l + 1/2)$ value is also the maximum value of the quantum number j. Since, as for any angular momentum, the different values of j must differ by one $\hbar$ unity, then we can conclude that for states with $l = 0$ we have $j = 1/2$, while for states with $l \geqslant 1$ we have $j = l \pm 1/2$.

We can now easily calculate the $\mathbf{S} \cdot \mathbf{L}$ product: since

$$J^2 = L^2 + S^2 + 2\,\mathbf{L} \cdot \mathbf{S} \tag{3.86}$$

we can write

$$\mathbf{L} \cdot \mathbf{S} = \frac{\hbar^2}{2}\ [j(j + 1) - l(l + 1) - s(s + 1)] \tag{3.87}$$

which eventually leads to

$$E_{so} = \langle \xi(r) \rangle\ \mathbf{S} \cdot \mathbf{L}$$
$$= \frac{\hbar^2}{2}\ \langle \xi(r) \rangle [j(j + 1) - l(l + 1) - s(s + 1)] \tag{3.88}$$

representing the most general expression for the quantum expectation value of the spin–orbit interaction energy in hydrogenic atoms.

In order to illustrate how the general result is applied in practice, let us consider a hydrogenic atom in the state with $n = 2$. The corresponding minimum and maximum values of the total angular momentum quantum number are $j = l - s$ and $j = l + s$, consistently with the formal properties defined above. The spin–orbit energy appearing in equation (3.88) does not affect the state with $n = 2$ and $l = 0$ and,

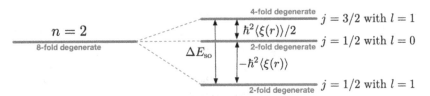

Figure 3.15. Spin–orbit splitting of the $n = 2$ state in an hydrogenic atom. The degeneracy of each level is explicitly reported, together with the relevant quantum numbers. A warning: a critical readdressing of this energy diagram will be presented when discussing figure 3.16.

therefore, we calculate them by setting $l = 1$ and $s = 1/2$ so as to obtain $j = 1/2$ or $3/2$ for which we get

$$E_{so}^{j=1/2} = -\hbar^2 \langle \xi(r) \rangle \quad \text{and} \quad E_{so}^{j=3/2} = \frac{1}{2} \hbar^2 \langle \xi(r) \rangle \tag{3.89}$$

This means that the eightfold degenerate $n = 2$ level[24] is split into three levels, as illustrated in figure 3.15: the level $(n, l) = (2, 0)$ is unaffected, while the levels $(n, l) = (2, 1)$ are resolved with a separation $\Delta E_{so} = 3\hbar^2 \langle \xi(r) \rangle/2$. The initial eight-fold degeneracy of the level is distributed as indicated in figure 3.15 by considering all possible values of the total magnetic quantum number m_j: in the topmost level we calculate the four possible values $m_j = \pm 1/2, \pm 3/2$ while in the intermediate and bottom levels we only have $m_j = \pm 1/2$.

3.5 The fine structure of one-electron atoms

3.5.1 Synopsis of all relativistic effects

Although we introduced the spin on the basis of experimental evidence, we already commented that its true origin is relativistic. As a matter of fact, spin is not the only relativistic feature in atomic physics. Although the development of a full quantum-relativistic theory falls beyond the scope of this primer and it can be found elsewhere [3, 4, 5, 8, 11], for sake of completeness we nevertheless provide a synopsis of all relativistic effects in hydrogenic atoms.

The relativistic wave equation fully accounting for special relativity in the context of quantum mechanics was derived in 1928 by P A M Dirac. For a single electron (we are considering hydrogenic atoms only) which interacts with an infinite-mass nucleus centred at the origin of the frame of reference, the Dirac equation provides three different relativistic corrections with respect to the Schrödinger (non-relativistic) results, namely: (i) a correction to the electron kinetic energy; (ii) the spin–orbit interaction term; and (iii) a correction to the electron potential energy. If Z is not too large, they all are very small and, therefore, the most practical way to evaluate them is to proceed by using perturbation theory. Since they are additive corrective terms, we can treat them separately.

---

[24] We must remember that, when considering the spin degree of freedom, a hydrogenic quantum state is $2n^2$-fold degenerate.

3.5.1.1 Relativistic correction to the electron kinetic energy

The non-relativistic kinetic energy $p^2/2m_e$ (where m_e is now more accurately understood as the electron *rest mass*) must be corrected by an additional term $-p^4/8m_e^3c^2$. This correction (which is valid to the order v^2/c^2, where v is the electron orbital speed) works as a perturbation to the spectrum of Schrödinger Hamiltonian operator given in equation (3.3) for which we know the (unperturbed) ψ_{nlm_l} eigenfunctions[25]. The corresponding energy correction is calculated as

$$\Delta E_{\text{rel}}^{\text{kin}} = \langle \psi_{nlm_l}|-\frac{p^4}{8m_e^3c^2}|\psi_{nlm_l}\rangle = -\frac{1}{2m_ec^2}\langle\psi_{nlm_l}|\hat{T}^2|\psi_{nlm_l}\rangle$$

$$= -E_n\frac{(Z\alpha)^2}{n^2}\left(\frac{3}{4} - \frac{n}{l+1/2}\right) \tag{3.90}$$

where $\hat{T}$ is the electron kinetic energy operator, E_n is given in equation (3.32) and

$$\alpha = \frac{e^2}{4\pi\varepsilon_0\hbar c} \simeq \frac{1}{137} \tag{3.91}$$

is referred to as the *fine structure constant*. Equation (3.90) provides evidence that the first relativistic correction is a pure shift of the Schrödinger levels.

3.5.1.2 The spin–orbit interaction term

This term has been extensively discussed in the previous section: it takes the E_{so} form given in equation (3.88) and also includes the Thomas precession correction.

In this case, the relativistic correction generates the splitting of levels with $l > 0$ which are degenerate in the Schrödinger treatment, as illustrated in figure 3.15.

3.5.1.3 Relativistic correction to the electron potential energy

A quantum-relativistic treatment of the electron physics proves that there exists a fundamental limitation on measuring its position which depends on the electron mass. More specifically, assuming that the electron is moving at speed c, its instantaneous position cannot be located more accurately than within a volume $\lambda_{\text{Compton}}^3 = (\hbar/m_ec)^3$; in other words, *the electron cannot be considered any longer as a point charge*. The quantity λ_{Compton} is known as *Compton wavelength*[26]: it is equal to the wavelength of a photon with the same energy as m_ec^2 (electron rest energy).

The consequence of this conclusion is important: the electron–nucleus Coulomb interaction must be weighted over the volume $\lambda_{\text{Compton}}^3$ and, therefore, the actual

---

[25] We remark that the Coulomb energy must be duly corrected by a factor Z when dealing with hydrogenic atoms. Similarly, hydrogenic eigenfunctions must be used for any atom with $Z > 1$ as discussed in section 3.2.
[26] It was introduced by A Compton in 1923 to explain the electron–photon scattering, usually referred to as Compton scattering [4, 5]. The Compton wavelength is equivalent to the de Broglie wavelength for particles with velocity $c/\sqrt{2}$: while this latter defines the threshold below which a particle should be treated as a wave, the Compton wavelength defines the threshold below which corrections to non-relativistic quantum mechanics become important.

situation does not exactly correspond to the case of two interacting point-like charges, as we assumed throughout the non-relativistic treatment. This causes a correction $(\pi\hbar^2/2m_e^2c^2)(Ze^2/4\pi\varepsilon_0)\delta(\mathbf{r})$ to the electron potential energy [3, 5] which, once again, we evaluate by means of the perturbation theory

$$
\begin{aligned}
\Delta E_{\text{rel}}^{\text{pot}} &= \frac{\pi\hbar^2}{2m_e^2c^2}\left(\frac{Ze^2}{4\pi\varepsilon_0}\right)\langle\psi_{nlm_l}|\delta(\mathbf{r})|\psi_{nlm_l}\rangle \\
&= \frac{\pi\hbar^2}{2m_e^2c^2}\left(\frac{Ze^2}{4\pi\varepsilon_0}\right)|\psi_{n00}(\mathbf{r}=0)|^2 \\
&= -E_n\frac{(Z\alpha)^2}{n}
\end{aligned}
\tag{3.92}
$$

where in computing the matrix element we have made explicit that $|\psi_{nlm_l}(\mathbf{r}=0)|^2$ is non-negligible only for s states (as we commented when discussing the radial part of the wavefunction in section 3.1.2). This relativistic energy contribution is referred to as the *Darwin correction*.

3.5.2 The fine structure spectrum

The overall relativistic correction $\Delta E_{\text{rel}}^{nj}$ to the energy levels obtained by solving the ordinary Schrödinger equation is obtained by summing the above three rectifications[27]

$$
\Delta E_{\text{rel}}^{nj} = \Delta E_{\text{rel}}^{\text{kin}} + E_{\text{so}} + \Delta E_{\text{rel}}^{\text{pot}} = E_n\frac{(Z\alpha)^2}{n^2}\left(\frac{n}{j+\frac{1}{2}}-\frac{3}{4}\right)
\tag{3.94}
$$

which finally leads to

$$
E_{nj} = E_n + \Delta E_{\text{rel}}^{nj} = E_n\left[1+\frac{(Z\alpha)^2}{n^2}\left(\frac{n}{j+\frac{1}{2}}-\frac{3}{4}\right)\right]
\tag{3.95}
$$

representing *the fine structure energy spectrum of hydrogenic atoms*. It is characterized by four main features: (i) relativistic hydrogenic energies depend on two quantum numbers, namely $n = 1, 2,...$ and $j = 1/2, 3/2,... , n - 1/2$; (ii) relativistic

---

[27] It is interesting to remark that a similar result is obtained in the framework of the Sommerfeld model for the hydrogen atom, discussed in section 1.3.2, by considering the relativistic corrections to kinetic energy, but not considering the spin degree of freedom at that time still unknown. Following this approach [3], it is calculated

$$
E_{\text{Sommerfeld}} = -\frac{m_eZ^2e^4}{(4\pi\varepsilon_0)^22n^2\hbar^2}\left[1+\frac{\alpha^2Z^2}{n}\left(\frac{1}{n_\theta}-\frac{3}{4n}\right)\right]
\tag{3.93}
$$

providing the fine spectrum of a Sommerfeld-like hydrogenic atom in terms of the two quantum numbers n and n_θ.

corrections are increasingly more important for heavier atoms; (iii) the three corrections are of the same order, but spin–orbit coupling dominates in large-Z atoms; and (iv) the $2n^2$-fold degeneracy is partially resolved. This latter feature is recognized by stating that *the relativistic corrections reveal the fine structure of the energy spectrum*, as reported in figure 3.16 in the specific case of the hydrogen atom. The actual shift for the ground state is easily calculated as

$$\Delta E_{\text{rel}}^{n=1,j=1/2} = (-13.6 \text{ eV}) \left(\frac{1}{137} \right)^2 \frac{1}{4} = -1.81 \times 10^{-4} \text{ eV} \qquad (3.96)$$

In figure 3.16 we have also indicated the degeneracy of each level: it is easily calculated by making use of the quantization rules holding for the **J** total angular momentum discussed in section 3.4. More specifically, if $j = 1/2$, 3/2, or 5/2 then there are, respectively, 2, 4 or 6 possible values of m_j.

We conclude this discussion with an important remark. By comparing figures 3.15 and 3.16 one might be confused at first glance: in fact, if we consider the $n = 2$ state, it is resolved into a triplet due to the spin–orbit interaction alone (see figure 3.15), whereas it appears in the form of a doublet when considering the complex of all relativistic effects (see figure 3.16). In fact, there is no contradiction, but it must be

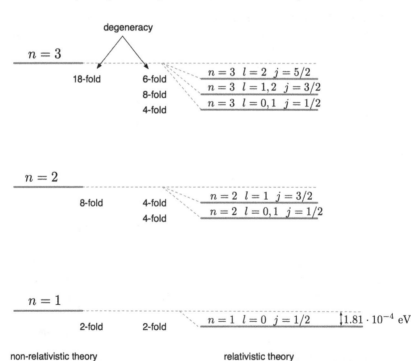

Figure 3.16. Fine structure of the hydrogen atom (right) compared to the non-relativistic structure provided by the Schrödinger equation (left). The degeneracy of each level is reported in the central column. Since the displacements of the relativistic levels from the Schrödinger ones are really very small (the actual value of the $n = 1$ state is explicitly reported), for illustration purposes the figure has been drawn not in scale.

made clear that figure 3.15 is only meant to pedagogically illustrate spin–orbit splitting, which, in that context, is the only effect in action: it follows that the energy of the levels depends on the quantum numbers (n, l, j). On the other hand, the more complex interplay among the full set of relativistic effects predicts through equation (3.95) that the energy actually depends only on the quantum numbers (n, j) and this makes the levels $2S_{1/2}$ and $2P_{1/2}$ merge again into a fourfold degenerate level. In summary: figure 3.15 is just the illustration of a model case, where only spin–orbit is at work, while the true physics is given by figure 3.16.

3.5.3 Lamb shift

According to equation (3.95) describing the fine structure of hydrogenic atoms, energy levels with same value of the j quantum number, but different values of the l quantum number should be degenerate. The reason is trivial: l does not explicitly appear in such an equation. Actually, this prediction of the theory does not correspond to the experimental results obtained by very high precision measurements: in 1947 it was in fact observed by W Lamb and R Retherford a very small (but still detectable) splitting between two levels which should instead coincide according to equation (3.95). They are defined by the quantum numbers $(l = 0, j = 1/2)$ and $(l = 1, j = 1/2)$ and both belong to the same $n = 2$ shell. The splitting takes the form of an upward shift of the $(l = 0, j = 1/2)$ level and, for this reason, it was henceforth referred to as the *Lamb shift*.

The quantum theory we are working with is not sophisticated enough to account for the Lamb shift, even including relativistic corrections. A much more refined theory is needed, performing the full quantum treatment of both the atomic system and the electromagnetic radiation field: this is the realm of *quantum electrodynamics* (QED), indeed a topic well beyond the level of this Primer[28]. It is, however, possible to provide a phenomenological argument able to justify the Lamb shift. QED proves that an atomic electron can virtually absorb and then emit a photon with energy $\Delta E = \hbar\omega$ without violating the energy conservation provided that the two processes occur over a time interval $\Delta t < \hbar/\Delta E$. This holds since the Heisenberg uncertainty principle dictates $\Delta E \Delta t \geqslant \hbar$. Virtual phenomena cause electron recoils, so that the electron motion results as random shaky. This implies that the electronic position $\mathbf{r}$ with respect to its atomic nucleus is smeared out by a factor $\delta\mathbf{r}$: the corresponding Coulomb interaction $V(r)$ is therefore modified with respect to the point-charge model, assuming the form $V_{\text{QED}}(r) = -(Ze^2/4\pi\varepsilon_0)1/(r + \delta r)$. We draw the conclusion that *the electron of a hydrogenic atom results as less strongly attracted by its nucleus because of this QED effect*. This very small correction is only relevant for $l = 0$ states since, as discussed in section 3.1.2, they can penetrate closer to the nuclear region than any other $l > 0$ state. This qualitatively explains why the $(l = 0, j = 1/2)$ level is Lamb-shifted upward.

The QED corrections to the fine-structure spectrum of the hydrogen atom are shown in figure 3.17(C).

---

[28] A tutorial (but still challenging) introduction to the QED description of atoms and their (self)interaction with radiation can be found in [12].

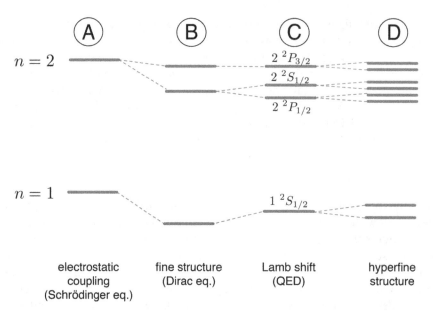

Figure 3.17. Complete energy spectrum (not in scale) of the hydrogen atom for the $n = 1$ and $n = 2$ states, reflecting four possible descriptions with increasing level of detail. (A) Only electrostatic coupling is considered, according to the Schrödinger equation; (B) spin–orbit and relativist effects are treated by the Dirac equation; (C) quantum electrodynamics (QED) is used for a quantum description of both the atomic system and the radiation field, producing the Lamb shift (the spectroscopic notation is used to label the states); (D) hyperfine structure taking into account nuclear effects.

3.5.4 Nuclear effects

The accuracy we reached in describing the energy spectrum of hydrogenic atoms is really very high, but it can be further improved. This requires describing nuclear effects beyond the level of erudition we developed so far. In particular, in appendix C we outline the treatment of the so-called *hyperfine structure* by considering explicitly the role played by the nuclear magnetic dipole moment. The ultimate energy spectrum of the hydrogen atom, including any effect so far discussed (Coulomb interactions, spin–orbit interactions, Lamb shift, and nuclear magnetic interactions) is reported in figure 3.17.

Finally, we remark that so far the nucleus has been treated as a point-like charge. Nuclear physics, instead, provides evidence that it has a small, but finite dimension of the order of $\sim 10^{-15}$ m [13]. Accordingly, we could take into consideration nucleus finite-size corrections as a perturbation. In appendix D we present their calculation showing that such corrections are indeed very small and, therefore, fully negligible.

3.5.5 Classifying the fine structure levels: the spectroscopic notation

It is useful to take full advantage from the existence of the three angular momenta **J**, **L**, and **S** to define *a classification scheme for the fine structure levels* [14] which is often called 'the spectroscopic notation' for the quantum states.

We agree to indicate a generic level by the symbol

$$n\ ^{2S+1}L_J \tag{3.97}$$

where n is the principal quantum number and L is the quantum number for the orbital angular momentum and it is represented by the corresponding letter, as discussed in section 3.1.4. The left superscript $2S + 1$ contains the spin quantum number S and it is usually referred to as the *spin multiplicity*. It is important to stress that the use of capital letters for the orbital and spin quantum numbers will be fully justified when discussing the physics of multi-electron atoms. We also remark that this notation is somewhat redundant for hydrogenic atoms containing just one electron since we always have $2S + 1 = 2$, but it will be very useful in the case of multi-electron atoms.

As shown in figure 3.17(C) the sequence of fine structure levels of the hydrogen atom is the following

$$1\ ^2S_{1/2},\ 2\ ^2P_{1/2},\ 2\ ^2S_{1/2},\ 2\ ^2P_{3/2},\ 3\ ^2P_{1/2},\ 3\ ^2S_{1/2},\ 3\ ^2P_{3/2},\ 3\ ^2D_{3/2},\ 3\ ^2D_{5/2},\ldots \tag{3.98}$$

This notation is used in the next section to rationalize the effects of an external magnetic field.

3.6 Anomalous Zeeman and Paschen–Back effects

In section 3.3.1 we studied the effects of a uniform magnetic field on the energy spectrum of hydrogenic atoms without considering the electronic spin. Now we want to complete the discussion by recognising that *the total electron magnetic moment is the sum of the orbital and the spin magnetic moments*

$$\mathbf{M}_{tot} = \mathbf{M}_L + \mathbf{M}_S = -\frac{\mu_B}{\hbar}\ (\mathbf{L} + 2\mathbf{S}) \tag{3.99}$$

where we made use of $g_L = 1$ and $g_S = 2$. While we will always assume to apply a static, homogeneous, external magnetic field $\mathbf{B} = (0, 0, B_z)$ to the atom, we can distinguish the two cases of weak and strong magnetic field, respectively, giving rise to the 'anomalous Zeeman effect' and 'Paschen–Back effect'. The common feature of both regimes is that, under the action of an external field, angular momenta undergoes precessional motions, as previously discussed in section 3.4. We will develop our theory within the 'vector model'[29].

In the weak field regime the spin–orbit coupling is stronger than the interaction between the electron magnetic dipoles and the external field: therefore, the precession

---

[29] A more rigorous analysis is based on the quantum mechanical perturbation theory. If the external field is smaller than the internal field due to the electron orbital motion, than its effects are expected to be smaller than the spin–orbit ones. Accordingly, the new potential energy term $E_{mag} = -\mathbf{M}_{tot} \cdot \mathbf{B}$ describing the coupling of the total electron magnetic moment with the external field can be treated as a perturbation: its effects are properly taken into account by calculating the expectation value of the corresponding quantum operator on the hydrogenic wavefunctions obtained by solving the Schrödinger equation where the Hamiltonian operator is given by equation (3.3). Such a non-trivial calculation requires determining the expectation value of the $\hat{\mathbf{J}} + \hat{\mathbf{S}}$ operator by means of the Wigner–Eckart theorem of the formal quantum theory for angular momenta [5, 6].

frequency of $\mathbf{L}$ and $\mathbf{S}$ around $\mathbf{J} = \mathbf{L} + \mathbf{S}$ is much higher than the precession frequency of $\mathbf{J}$ around $\mathbf{B}$. This implies that the contributions of the orbital and spin angular momenta to the corresponding magnetic dipoles are provided by *their average values, calculated over a precessional period*. The components normal to $\mathbf{J}$ average to zero because of the quick precession and, therefore, the only active components are those along $\mathbf{J}$

$$\mathbf{L}_{\text{ave}} = \left(\mathbf{L} \cdot \frac{\mathbf{J}}{J} \right) \frac{\mathbf{J}}{J} = \frac{(\mathbf{L} \cdot \mathbf{J})}{J^2} \, \mathbf{J}$$
$$\mathbf{S}_{\text{ave}} = \frac{(\mathbf{S} \cdot \mathbf{J})}{J^2} \, \mathbf{J}$$

(3.100)

The resulting effective orbital and spin magnetic moments are easily computed as

$$\mathbf{M}_{L,\text{ave}} = - \frac{\mu_B}{\hbar} \frac{(\mathbf{L} \cdot \mathbf{J})}{J^2} \, \mathbf{J}$$
$$\mathbf{M}_{S,\text{ave}} = - 2 \frac{\mu_B}{\hbar} \frac{(\mathbf{S} \cdot \mathbf{J})}{J^2} \, \mathbf{J}$$

(3.101)

and this implies that

$$\begin{aligned}
\mathbf{M}_{\text{tot,ave}} &= \mathbf{M}_{L,\text{ave}} + \mathbf{M}_{S,\text{ave}} \\
&= - \frac{\mu_B}{\hbar} \frac{1}{J^2} [(\mathbf{L} \cdot \mathbf{J}) + 2(\mathbf{S} \cdot \mathbf{J})] \, \mathbf{J} \\
&= - \frac{\mu_B}{\hbar} \frac{1}{J^2} [(\mathbf{J} + \mathbf{S}) \cdot \mathbf{J}] \, \mathbf{J} \\
&= - \frac{\mu_B}{\hbar} \frac{1}{J^2} \left[J^2 + \frac{J^2 - L^2 + S^2}{2} \right] \mathbf{J}
\end{aligned}$$

(3.102)

where we made use of the relations $\mathbf{L} + 2\mathbf{S} = \mathbf{J} + \mathbf{S}$ and $\mathbf{J} \cdot \mathbf{S} = 1/2(J^2 - L^2 + S^2)$. The quantum expectation value of $\mathbf{M}_{\text{tot,ave}}$ is easily calculated since we already know the expectation value of any angular momentum appearing in equation (3.102)

$$\langle \mathbf{M}_{\text{tot,ave}} \rangle = - \frac{\mu_B}{\hbar} g_j \, \mathbf{J}$$

(3.103)

which is written in the same form used in equation (3.44) and in equation (3.53) thanks to the definition of the *Landé g-factor*

$$g_j = 1 + \frac{j(j+1) - l(l+1) + s(s+1)}{2j(j+1)}$$

(3.104)

The coupling between $\mathbf{M}_{\text{tot,ave}}$ and the external field $\mathbf{B}$ affects the fine structure spectrum given in equation (3.95) by the addition a new potential energy term

$$E_{\text{mag}} = - \langle \mathbf{M}_{\text{tot,ave}} \rangle \cdot \mathbf{B} = \mu_B \, g_j \, m_j \, B_z$$

(3.105)

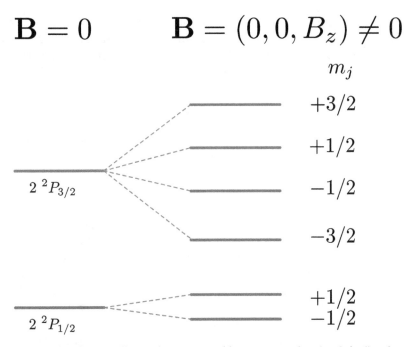

Figure 3.18. Anomalous Zeeman effect on the two states with quantum numbers ($n = 2$, $l = 1$) and, respectively, $j = 1/2$ (spectroscopic notation $2^2 P_{1/2}$) and $j = 3/2$ (spectroscopic notation $2^2 P_{3/2}$) of the hydrogen atom (hyperfine structure neglected for sake of clarity). Note the different amplitude of the Zeeman splitting due to the different values of the Landé g-factor.

which causes the splitting of levels with different quantum number m_j, as reported in figure 3.18. We remark that the Landé g-factor depends on both the l and j quantum numbers[30] and, therefore, *the magnitude of the anomalous Zeeman splitting depends on the specific level considered*: this is at variance with the result found for the normal Zeeman effect, where any pair of levels was found to split by the very same $\Delta E_{\mathrm{mag}} = \mu_{\mathrm{B}} B_z$ amount. This is the reason for using the word 'anomalous': when taking into account the spin degree of freedom, the situation is more complicated with respect to the discussion reported in section 3.3.1.

Let us now turn to the opposite limiting case of a *strong external field regime: now the coupling of* **L** *and* **S** *with the magnetic field* **B** *overcomes spin–orbit effects* and, therefore, *both angular momenta undergo precessional motions around the field direction*. This is known as Paschen–Back effect. In this case the spin–orbit interaction can be neglected in the first instance. Accordingly, the hydrogenic energy spectrum is affected by the addition of the new potential energy term

$$E_{\mathrm{mag}} = -\mathbf{M}_L \cdot \mathbf{B} - \mathbf{M}_S \cdot \mathbf{B} = \mu_{\mathrm{B}} \left(m_l + 2m_s \right) B_z \tag{3.106}$$

---

[30] In one-electron atoms of course it always holds $s(s + 1) = 3/4$.

where we made explicit the fact that in this regime each magnetic moment is singularly coupled to the magnetic field. In order to eventually add the spin–orbit fine structure term we observe that the **L** and **S** components normal to the field average to zero. Therefore, the spin–orbit energy given in equation (3.88) assumes the simplified form

$$E_{so}^{PB} = \langle \xi(r) \rangle \mathbf{L} \cdot \mathbf{S} = \langle \xi(r) \rangle L_z S_z = \langle \xi(r) \rangle \hbar^2 m_l m_s \tag{3.107}$$

where the superscript 'PB' simply reminds us that this expression is valid only for the situation corresponding to the Paschen–Back effect.

3.7 The action of an electric field

The energy spectrum of an atom is also modified by the application of an external electric field **E**, which we assume is static and homogeneous. The resulting modification, called *Stark effect*, is here discussed in the specific case of the hydrogen atom without considering relativistic effects. In other words, the unperturbed energy spectrum corresponds to figure 3.5. By assuming that the electric field is weak enough, we can treat its effects as perturbations, according to the formalism developed in section 2.7.2. If we further set $\mathbf{E} = (0, 0, E_z)$, then its effect on the hydrogen atom is properly taken into account by adding a new potential energy term

$$E_{elec} = -(-e\mathbf{r}) \cdot \mathbf{E} = e \, E_z \, z \tag{3.108}$$

to the unperturbed levels of equation (3.32), where **r** is the electron position (see figure 3.1) and $-e\mathbf{r}$ is its *electric dipole moment*. This classical expression corresponds to the perturbation operator

$$\hat{H}_{pert} = e \, E_z \, z \tag{3.109}$$

whose effects are calculated by using equation (2.96).

Let us consider the $1s$ hydrogen ground-state characterized by the wavefunction ψ_{1s}[31]. The first-order correction $\Delta E_{1s}^{(1)}$ is calculated according to equation (2.93)

$$\Delta E_{1s}^{(1)} = \int \psi_{1s}^* \hat{H}_{pert} \psi_{1s}^* d\mathbf{r} = eE_z \int \psi_{1s}^* z \, \psi_{1s} \, d\mathbf{r} = 0 \tag{3.110}$$

where the superscript (1) properly indicates that this is the first-order correction. This result is a direct consequence of a twofold mathematical feature, namely (i) the ψ_{1s}^* and ψ_{1s} wavefunctions have the same parity, and (ii) the z term has odd parity. We conclude that *for the hydrogen atom there is no first-order Stark effect*.

We now consider second-order effect $\Delta E_{1s}^{(2)}$. By using equation (2.96), in this case we have

---

[31] There is no need to take care of the spin term, since the perturbation only acts on the space degrees of freedom: spin is not active in the Stark effect.

$$\Delta E_{1s}^{(2)} = e^2 E_z^2 \sum_{n,l,m_l \neq 1,0,0} \frac{\left| \int \psi_{nlm_l}^* \, z \, \psi_{1s} \, d\mathbf{r} \right|^2}{E_{1s} - E_n} \neq 0 \tag{3.111}$$

where E_{1s} and E_n (with $n \neq 1$) are the unperturbed ground-state and excited-states energies, respectively. In conclusion, *the Stark effect on the hydrogen ground-state is a second-order effect*. The actual calculation of the sum appearing on the right-hand side of this equation is not an easy task (it can be found elsewhere [3]), eventually leading to

$$\Delta E_{1s}^{(2)} = -\frac{9}{4} (4\pi\varepsilon_0) \, a_0^3 \, E_z^2 \tag{3.112}$$

For typical values of the external electric field $E_z \sim 10^7$–10^9 V m^{-1} the Stark energy correction is indeed very small.

If we now pass to consider excited states, we remark that *for them the Stark effect is effective even at the first-order*. For instance, it is proved [3] that for the $n = 2$ state the fourfold degeneracy is partially removed since the two levels with $m_l = 0$ are upward/downward shifted by the amount $\pm 3ea_0 E_z$ with respect to the unperturbed $E_{n=2}$ level. The situation is illustrated in figure 3.19.

By means of elementary electrostatics, we can define the *induced electric dipole moment* for the 1s state as

$$d_{\text{elec}}^{1s} = -\frac{d\Delta E_{1s}^{(2)}}{dE_z} = \alpha_{\text{elec}}^{1s} \, E_z \tag{3.113}$$

where $\alpha_{\text{elec}}^{1s}$ is the *electric dipole polarisability* for this quantum state

$$\alpha_{\text{elec}}^{1s} = \frac{9}{2} (4\pi\varepsilon_0) \, a_0^3 \tag{3.114}$$

providing a direct way to estimate the distortion of the electron cloud operated by the external field. In summary, the net Stark effect on the hydrogen 1s state consists (i) in a very small energy lowering quantified by equation (3.112) and (ii) in a distortion of the electron cloud.

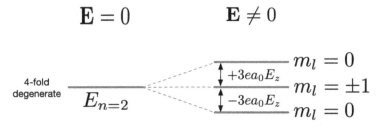

Figure 3.19. Linear Stark effect on the $n = 2$ state.

Further reading and references

[1] Jackson J D 1975 *Classical Electromagnetism* (New York: Wiley)

[2] Feynman R P, Leighton R B and Sands M 1963 *The Feynman Lectures on Physics* (Reading, MA: Addison-Wesley)

[3] Bransden B H and Joachain C J 1983 *Physics of Atoms and Molecules* (Harlow: Addison-Wesley Longman)

[4] Eisberg R and Resnick R 1985 *Quantum Physics of Atoms, Molecules, Solids, Nuclei, and Particles* 2nd edn (Hoboken, NJ: Wiley)

[5] Demtröder W 2010 *Atoms, Molecules and Photons* (Heidelberg: Springer)

[6] Sakurai J J and Napolitano J 2011 *Modern Quantum Mechanics* 2nd edn (Reading, MA: Addison-Wesley)

[7] Griffiths D J and Schroeter D F 2018 *Introduction to Quantum Mechanics* 3rd edn (Cambridge: Cambridge University Press)

[8] Greiner W 1990 *Relativistic Quantum Mechanics* (Heidelberg: Springer)

[9] Goldstein H 1996 *Classical Mechanics* 2nd edn (Reading, MA: Addison-Wesley)

[10] Leighton R B 1959 *Principles of Modern Physics* (New York: McGraw-Hill)

[11] Bransden B H and Joachain C J 2000 *Quantum Mechanics* (Englewood Cliffs, NJ: Prentice-Hall)

[12] Hestenes D 1983 Quantum mechanics from self-interaction *Found. Phys.* **15** 63

[13] Povh B, Rith K, Scholz C and Zetsche F 2009 *Particles and Nuclei* 6th edn (Heidelberg: Springer)

[14] Foot C J 2005 *Atomic Physics* (Oxford: Oxford University Press)

IOP Publishing

Atomic and Molecular Physics (Second Edition)
A primer
Luciano Colombo

Chapter 4

Interaction of one-electron atoms with radiation

Syllabus—*We treat the interaction between an electromagnetic radiation bath and a material system in the semi-classical picture: the radiation is treated according to the Maxwell equations, but the process of emission/absorption is described through the concept of a photon; on the other hand, the material system is modelled according to quantum mechanics, assuming a discrete two-level energy spectrum. This very simple model allows one to elaborate the elementary theory of the absorption/emission coefficients, both for the spontaneous and for the stimulated processes which will be described within the electric dipole approximation. Selection rules for allowed transitions are eventually obtained and the nature of high-order transition is briefly outlined. Finally, a qualitative description of the light amplification by stimulated emission of radiation (LASER) is presented.*

4.1 Emission and absorption of radiation

4.1.1 Problem definition and some useful approximations

The interaction of a hydrogenic atom with static electric or magnetic fields has been extensively discussed in chapter 3: we have learned that their effects basically consist in shifting and/or splitting the energy levels of the unperturbed system. Now we want to extend our knowledge to the effects induced by an electromagnetic (e.m.) radiation, namely to the case in which the electric and magnetic fields vary in time.

The most rigorous theoretical framework to address this problem is quantum electrodynamics, where both the atomic system and the radiation field are treated quantum mechanically [1]. This treatment, however, falls far beyond the scope of this primer and, therefore, we will rely on a *semi-classical picture* according to which the atomic system is described quantum mechanically, while the e.m. radiation is discussed as a classical field fulfilling Maxwell equations [2]. We will also assume that *the intensity of the radiation is not too high*. This implies that its effects can be considered as perturbations on the spectrum of the isolated atom. More specifically, we will assume that *the energy spectrum of the atom remains unaffected by the radiation field, whose net effect will only be to promote electronic transitions between stationary states.* These transitions will take place between discrete energy levels and, for simplicity, we will neglect relativistic effects. Despite these approximations, the resulting model will be accurate enough to catch the main features underlying the mechanisms of absorption and emission of light by an atomic system.

4.1.2 Emission and absorption

Let us consider two atomic levels with energy E_1 and E_2, respectively described by the hydrogenic wavefunctions ψ_1 and ψ_2. If we further assume that $E_1 < E_2$ we will refer to ψ_2 as the 'excited state', while ψ_1 will be named the 'ground state'[1].

We begin our discussion by considering the case of an *isolated atom*, i.e., the case in which *no radiation field is present*. If the atom initially occupies the excited state with energy E_2, then a spontaneous decay $\psi_2 \rightarrow \psi_1$ is observed, accompanied by the emission of a photon with frequency $\nu_{21} = |E_1 - E_2|/h$. This process is usually referred to as *spontaneous emission*. The situation is sketched in figure 4.1. We must duly remark that, strictly speaking, within our semi-classical model such a spontaneous emission should not occur, since ψ_2 is a stationary state of the time-independent Hamiltonian operator describing the atom (see discussion in section 2.4.2). On the other hand, within a quantum electrodynamics treatment it is proved that ψ_2 is not a stationary state of the full Hamiltonian operator describing both the atom and the radiation field. This fundamental issue is translated into our semi-classical model by admitting that *the excited state-although stationary- may undergo a transition through a photon emission.*

On the other hand, whenever the atom is subjected to the action of an e.m. field two different processes may occur: (i) if it initially occupies the ground state, the transition $\psi_1 \rightarrow \psi_2$ is in fact observed, due to the absorption by the atom of a photon with frequency $\nu_{12} = |E_2 - E_1|/h$ subtracted from the radiation bath; (ii) in the

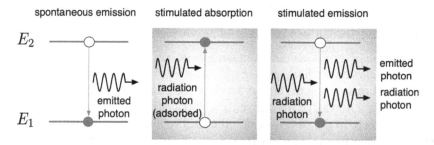

Figure 4.1. Schematic representation of the spontaneous emission (left), stimulated absorption (middle), and stimulated emission (right) of a photon by a two-state quantum system. In the case of stimulated processes, the presence of a radiation bath is rendered by the shadowing.

---

[1] We remark that in this context the ground state is not necessarily the lowest-energy one of the hydrogenic spectrum. Rather, it is just the state with comparatively lower energy between the two considered, i.e., the state which is eventually reached by the atom upon a suitable energy loss.

opposite situation where the state initially occupied is the excited one, the transition $\psi_2 \rightarrow \psi_1$ is observed accompanied by the emission of a photon with frequency $\nu_{21} = |E_1 - E_2|/h$. The two processes are called *stimulated absorption* and *stimulated emission*, respectively. The word 'stimulated' means that such processes are activated by the radiation field.

We finally remark that the absorption frequency ν_{12} and the emission frequency ν_{21} (for both spontaneous and stimulated emission) are just the same, according to the phenomenological discussion reported in section 1.2.2. Furthermore, stimulated phenomena are expected to depend on the spectral energy density u_ν of the e.m. radiation at the frequency $\nu = \nu_{12} = \nu_{21}$.

4.1.3 Einstein coefficients

It is customary to describe emission/absorption processes by their *transition rates* which characterize the evolution of the atomic populations occupying the different states involved in the transitions.

Let us define $A_{21}^{(\nu)}$ as the probability per unit time to observe the transition $\psi_2 \rightarrow \psi_1$ by spontaneous emission. If N_2 is the number of atoms in the state ψ_2, then its variation in time is given by

$$\dot{N_2} = -A_{21}^{(\nu)} N_2 \tag{4.1}$$

where $\dot{N_2} = dN_2/dt$; this simple differential equation is immediately solved as

$$N_2 = N_2^{(0)} \exp[-t/\tau_2] \tag{4.2}$$

where $N_2^{(0)}$ is the number of atoms initially occupying the state and $\tau_2 = 1/A_{21}^{(\nu)}$ is the *average lifetime of the state* ψ_2 which, as more extensively discussed in section 4.5 (see also [3]), is related to the finite width of the spectroscopic line associated with the emission transition.

For stimulated processes, the occurrence of any transition actually depends on the available number of photons with suitable energy or, equivalently, on the spectral energy density u_ν of the radiation field. Therefore, the probability per unit time to observe the stimulated absorption transition $\psi_1 \rightarrow \psi_2$ is cast in the form $u_{\nu_{12}} B_{12}^{(\nu)}$. Similarly, for the stimulated emission $\psi_2 \rightarrow \psi_1$ we set $u_{\nu_{21}} B_{21}^{(\nu)}$ its probability of occurrence per unit time.

The $A_{21}^{(\nu)}$, $B_{12}^{(\nu)}$ and $B_{21}^{(\nu)}$ terms are known as *Einstein coefficients* (in the frequency representation, as detailed below)[2]: their knowledge enables performing a full analysis of the populations when absorption and emission phenomena are simultaneously active, as discussed in the next section.

---

[2] We remark that, according to their true definition, spontaneous and stimulated coefficients are measured in different units.

4.1.4 Population analysis

We now consider a population of many identical atoms subjected to a radiation field with spectral energy density u_ν. Let N_1 and N_2 be the number of atoms respectively occupying the ground state '1' and the excited state '2', while $N_1^{(0)}$ and $N_2^{(0)}$ are their populations at time zero. We want to predict how these numbers vary in time because of absorption/emission phenomena.

We indicate by $dT_{1\to2}$ and $dT_{2\to1}$ the number of absorption and emission transitions, respectively, occurring in the infinitesimal time dt. It is straightforward to write

$$\begin{cases} dT_{1\to2} = N_1 \, B_{12}^{(\nu)} u_{\nu_{12}} dt \\ dT_{2\to1} = N_2 \, [B_{21}^{(\nu)} u_{\nu_{21}} + A_{21}^{(\nu)}] dt \end{cases} \qquad (4.3)$$

as the total balance among all the allowed phenomena. If the total [atoms + radiation field] system is in thermodynamical equilibrium at temperature T, then it holds $dT_{1\to2} = dT_{2\to1}$ so that

$$\frac{N_1}{N_2} \, B_{12}^{(\nu)} u_{\nu_{12}}^{BB} = B_{21}^{(\nu)} u_{\nu_{21}}^{BB} + A_{21}^{(\nu)} \qquad (4.4)$$

where we have also made it clear that the radiation field is described by a blackbody energy density u^{BB} because of the assumed equilibrium condition. By the Boltzmann distribution[3] we easily get

$$\frac{N_1}{N_2} = \exp[(E_2 - E_1)/k_B T] = \exp[h\nu_{21}/k_B T] \qquad (4.5)$$

and using equation (1.14) we set

$$u_{\nu_{12}}^{BB} = u_{\nu_{21}}^{BB} = \frac{8\pi h}{c^3} \frac{\nu_{12}^3}{\exp[h\nu_{12}/k_B T] - 1} \qquad (4.6)$$

so that the combination of equations (4.4), (4.5), and (4.6) leads to

$$B_{12}^{(\nu)} = B_{21}^{(\nu)} \qquad (4.7)$$

$$A_{21}^{(\nu)} = \frac{8\pi h}{c^3} \nu_{12}^3 \, B_{21}^{(\nu)} \qquad (4.8)$$

---

[3] The Boltzmann distribution provides the number $N(E)$ of atoms with energy E when the total system is in equilibrium at temperature T. More specifically, we have: $N(E) = (N_{tot}/\mathcal{Z})\exp[-E/k_B T]$, where N_{tot} is the total number of atoms, k_B is the Boltzmann constant and $\mathcal{Z}$ is the system partition function [4] (we are neglecting the possibility that states may have different intrinsic occupation probabilities). A more thorough presentation of this topic is reported in appendix E.

a very remarkable result which is straightforwardly interpreted as follows:

- *stimulated emission and stimulated absorption are physically equivalent phenomena*, since they are described by identical transition rates[4];
- *the larger the energy difference between the ground and excited states, the more intense is the spontaneous emission* (governed by the $A_{21}^{(\nu)}$ coefficient) with respect to the stimulated one (governed by the B_{21} coefficient).

It is finally interesting to attempt the *energy balance* for the population of atoms collectively undergoing absorption/emission phenomena. Still considering an equilibrium situation between the radiation bath and the material system, the total *absorbed* energy and the total *emitted* energy are $N_1^{(0)} B_{12}^{(\nu)} u_{\nu_{12}}^{BB} h\nu_{12}$ and $N_2^{(0)} \left[B_{21}^{(\nu)} u_{\nu_{21}}^{BB} + A_{21}^{(\nu)} \right] h\nu_{12}$, respectively. Sometimes the relation $B_{12}^{(\nu)} = B_{21}^{(\nu)}$ is referred to as the *detailed balance*.

4.1.5 Another representation

The spectral density of the blackbody radiation can also be expressed in the representation of the angular frequencies $\omega = 2\pi\nu$

$$u_\omega^{BB} = \frac{\hbar}{\pi^2 c^3} \frac{\omega^3}{\exp[\hbar\omega/k_B T] - 1} \tag{4.9}$$

nevertheless taking into account that

$$u_\nu^{BB} d\nu = u_\omega^{BB} d\omega \tag{4.10}$$

since the number of electromagnetic modes per unitary spectral range does not obviously vary depending on whether 'frequency' or 'angular frequency' is used. This implies that

$$u_\omega^{BB} = \frac{d\nu}{d\omega} u_\nu^{BB} = \frac{1}{2\pi} u_\nu^{BB} \tag{4.11}$$

where a conversion $1/2\pi$ term must be carefully taken into account whenever switching from one representation to the other.

The probability per unit time to observe the stimulated absorption transition undergoes a similar conversion

$$B_{12}^{(\nu)} u_\nu^{BB} = B_{12}^{(\omega)} u_\omega^{BB} \tag{4.12}$$

leading to

$$B_{12}^{(\omega)} = 2\pi B_{12}^{(\nu)} \tag{4.13}$$

representing a subtle, but important, notational specification.

---

[4] This result is confirmed by the more accurate quantum electrodynamics theory [1].

4.2 Microscopic theory of Einstein coefficients

The complete calculation of the microscopic expressions for the Einstein coefficients is rather complex and consists of a number of different steps, which will be dealt with separately.

4.2.1 Minimal coupling scheme

In order to investigate the effects of the perturbation operated by the e.m. field on the atomic system, we need to better define the underlying theoretical framework. We choose to describe the radiation field in the *Coulomb gauge* [2] where the electrostatic potential V and the vector potential $\mathbf{A}$ are set as

$$V = 0 \quad \nabla \cdot \mathbf{A} = 0 \tag{4.14}$$

and we furthermore assume the *harmonic time dependence*

$$\mathbf{A}(\mathbf{r}, t) = \mathbf{A} \exp\left[i(\mathbf{k} \cdot \mathbf{r} - \omega t)\right] \tag{4.15}$$

where $\mathbf{k}$ is the wavevector of the monochromatic e.m. wave of frequency ω. By expanding the $\exp(i\mathbf{k} \cdot \mathbf{r})$ term in a power series we get

$$\mathbf{A}(\mathbf{r}, t) = \mathbf{A} \exp(-i\omega t)\left[1 + i(\mathbf{k} \cdot \mathbf{r}) - \frac{1}{2}(\mathbf{k} \cdot \mathbf{r})^2 + \cdots \right] \simeq \mathbf{A} \exp(-i\omega t) \tag{4.16}$$

where we have restricted the expansion just to the first-order term. This holds provided that $\mathbf{k} \cdot \mathbf{r} \ll 1$, as is indeed the case: since $\mathbf{r}$ describes the spatial extension of the material system (i.e., the atom), this condition applies to any e.m. radiation, with the sole exclusion of those falling in the spectral range of x-rays and γ-rays. In other words, the condition $\mathbf{k} \cdot \mathbf{r} \ll 1$ actually applies to any radiation wavelength $\lambda = 2\pi|\mathbf{k}|^{-1}$ falling in the spectral range normally adopted in spectroscopic measurements. Therefore, with the sole exception of x- and γ-rays, the development of equation (4.16) is legitimate for all of the applications we are interested in.

In section 4.1.1 we anticipated that relativistic effects will be disregarded in our attempt to develop a quantum model for the interaction between an atom and a radiation field. Therefore, according to standard quantum mechanics [5, 6] we can describe the action of the e.m. field on the spinless electron of a hydrogenic atom through the so-called '*minimal coupling*' scheme, consisting in the following replacement of the kinetic energy operator

$$\underbrace{-\frac{\hbar^2}{2m_e} \nabla^2}_{\text{no e.m. field}} \quad \rightarrow \quad \underbrace{\frac{1}{2m_e}(-i\hbar\nabla + e\,\mathbf{A})^2}_{\text{with e.m. field}} \tag{4.17}$$

By developing the term in round brackets and, consistently with the initial assumption of weak field, by neglecting the quadratic term in the potential vector, it is easily shown that

$$\frac{1}{2m_e}(-i\hbar\nabla + e\,\mathbf{A})^2 \simeq -\frac{\hbar^2}{2m_e}\nabla^2 - \frac{ie\hbar}{m_e}\,\mathbf{A} \cdot \nabla \tag{4.18}$$

as thoroughly described in appendix F, where the subtle quantization procedure is developed in detail.

The result given in equation (4.18) has a truly transparent physical interpretation: (i) the first term on the right-hand-side of the above equation represents the ordinary electron kinetic energy operator; (ii) the second term represents the overall resulting effect of the e.m. radiation on the atomic system and, therefore, it corresponds to the operator describing the perturbation set on by the field. More specifically, since the vector potential is given in equation (4.16), we conclude that *the action of the electromagnetic field is that of a time-dependent perturbation to be treated as discussed in section 2.7.3.* There we learned that a quantum system with a discrete energy spectrum subjected to a time-dependent perturbation undergoes transitions between states at a rate provided by the Fermi golden rule given in equation (2.114). In the specific case we are discussing, the perturbation operator is

$$\hat{W} = -\frac{ie\hbar}{m_e}\mathbf{A} \cdot \nabla \qquad (4.19)$$

so that

$$\mathcal{P}_{1 \to 2} = \frac{2\pi}{\hbar} \left| \int \psi_2^* \left(-\frac{ie\hbar}{m_e} \mathbf{A} \cdot \vec{\nabla} \right) \psi_1 \, d\mathbf{r} \right|^2 \delta(|E_2 - E_1| - \hbar\omega) \qquad (4.20)$$

provides the *probability per unit time for a stimulated absorption transition* $\psi_1 \to \psi_2$.

4.2.2 Transition rates

The explicit calculation of equation (4.20) is a bit laborious and, therefore, it is convenient to adopt some formal shortcuts. First of all, it is convenient to rewrite equation (4.16) in the form

$$\mathbf{A}(\mathbf{r}, t) = A \, \mathbf{n} \, \exp(-i\omega t) \qquad (4.21)$$

where $\mathbf{n}$ is the polarization vector of the e.m. wave so that equation (4.20) becomes

$$\mathcal{P}_{1 \to 2} = \frac{2\pi A^2}{\hbar m_e^2} \left| \int \psi_2^* [e\mathbf{n} \cdot (-i\hbar\vec{\nabla})]\psi_1 d\mathbf{r} \right|^2 \delta(|E_2 - E_1| - \hbar\omega) \qquad (4.22)$$

a very convenient formulation for the calculations we are going to develop.

Let us start calculating the matrix element appearing in equation (4.22)

$$\int \psi_2^* [e\mathbf{n} \cdot (-i\hbar\vec{\nabla})]\psi_1 d\mathbf{r} = e \, \mathbf{n} \cdot \int \psi_2^* (-i\hbar\vec{\nabla})\psi_1 d\mathbf{r}$$

$$= e \, \mathbf{n} \cdot \int \psi_2^* \hat{\mathbf{p}} \psi_1 d\mathbf{r} \qquad (4.23)$$

$$= e \, m_e \, \mathbf{n} \cdot \int \psi_2^* \hat{v} \psi_1 d\mathbf{r}$$

where we used the momentum $\hat{\mathbf{p}} = -i\hbar\vec{\nabla}$ and velocity $\hat{v} = \hat{\mathbf{p}}/m_e$ operators. The matrix element of the velocity operator can be easily calculated by using the $[\hat{\mathbf{r}}, \hat{H}] = i\hbar\hat{v}$ commuting rule [5, 6]

$$
\begin{aligned}
\int \psi_2^* \hat{v}\psi_1 d\mathbf{r} &= \frac{1}{i\hbar} \int \psi_2^*[\hat{\mathbf{r}}, \hat{H}]\psi_1 d\mathbf{r} \\
&= \frac{1}{i\hbar}(E_2 - E_1) \int \psi_2^*(-\hat{\mathbf{r}})\psi_1 d\mathbf{r} \\
&= \frac{1}{i}\omega_{12} \int \psi_2^*(-\hat{\mathbf{r}})\psi_1 d\mathbf{r}
\end{aligned}
\tag{4.24}
$$

where it is understood that $\omega_{12} = (E_2 - E_1)/\hbar$, while $\hat{H}$ is energy operator in the absence of the e.m. radiation field (unperturbed Hamiltonian). Using equation (4.24) to calculate equation (4.22) eventually leads to

$$
\begin{aligned}
\mathcal{P}_{1\to2} &= \frac{2\pi A^2 \omega_{12}^2}{\hbar} \left| \int \psi_2^*[\mathbf{n} \cdot (-e\hat{\mathbf{r}})]\psi_1 d\mathbf{r} \right|^2 \delta(|E_2 - E_1| - \hbar\omega) \\
&= \frac{2\pi A^2 \omega_{12}^2}{\hbar} \cos^2\theta \left| \int \psi_2^*(-e\mathbf{r})\psi_1 d\mathbf{r} \right|^2 \delta(|E_2 - E_1| - \hbar\omega)
\end{aligned}
\tag{4.25}
$$

where θ is the angle between the polarization vector $\mathbf{n}$ and the electric dipole moment[5] $\mathbf{d} = -e\mathbf{r}$. Considering a non-polarized and isotropic radiation[6], we can replace $\cos^2\theta$ with its mean value[7] so that

$$
\begin{aligned}
\mathcal{P}_{1\to2} &= \frac{2\pi A^2 \omega_{12}^2}{3\hbar} \left| \int \psi_2^*(-e\mathbf{r})\psi_1 d\mathbf{r} \right|^2 \delta(|E_2 - E_1| - \hbar\omega) \\
&= \frac{2\pi A^2 \omega_{12}^2}{3\hbar^2} \left| \int \psi_2^*(-e\mathbf{r})|\psi_1 d\mathbf{r} \right|^2 \delta(|\omega_2 - \omega_1| - \omega)
\end{aligned}
\tag{4.26}
$$

where

$$
\begin{aligned}
& \left| \int \psi_2^*(-e\mathbf{r})\psi_1 d\mathbf{r} \right|^2 \\
&= \left| \int \psi_2^*(-ex)\psi_1 d\mathbf{r} \right|^2 + \left| \int \psi_2^*(-ey)\psi_1 d\mathbf{r} \right|^2 + \left| \int \psi_2^*(-ez)|\psi_1 d\mathbf{r} \right|^2
\end{aligned}
\tag{4.27}
$$

since $\mathbf{r} = (x, y, z)$. Thus, in an explicit calculation, the evaluation of the matrix element of electric dipole moment actually involves three separate calculations, one for each Cartesian component.

---

[5] In this step we made it explicit the fact that the quantum operator associated with the position is just the position vector itself, as reported in table 2.1.
[6] This is tantamount to guessing that $\mathbf{n}$ is randomly oriented.
[7] By averaging over the solid angle it is calculated that $(\cos^2\theta)_{\text{mean}} = \dfrac{\int_0^{2\pi} d\phi \int_0^{\pi} d\theta \, \cos^2\theta \sin\theta}{\int_0^{2\pi} d\phi \int_0^{\pi} d\theta \, \sin\theta} = \dfrac{1}{3}$

4.2.3 Intensity of the radiation field

The intensity A^2 of the e.m. radiation field appearing in equation (4.26) can be equivalently calculated according to classical electromagnetism or by using the quantum notion of a photon. Let us start with the classical theory where

$$u_\omega^{\text{classical}} = 2\varepsilon_0\omega^2 A^2 \tag{4.28}$$

provides the spectral energy density in the representation of the angular frequency [2]. On the other hand, if we name $N(\omega)$ the number of photons with energy $\hbar\omega$, we can write their spectral energy density as

$$u_\omega^{\text{quantum}} = \frac{N(\omega)\hbar\omega}{V} \tag{4.29}$$

where V is the space volume where the radiation field is contained. Of course, the classical and quantum picture must provide the same density: by setting $u_\omega^{\text{classical}} = u_\omega^{\text{quantum}}$ we immediately obtain

$$A^2(\omega) = \frac{N(\omega)\hbar}{2\varepsilon_0\omega V} \tag{4.30}$$

to be used in equation (4.26).

4.2.4 Stimulated absorption and stimulated emission coefficients

We can finally calculate the stimulated absorption coefficient $B_{12}^{(\omega)}$ in the representation of the angular frequency by using the definition given in section 4.1.3

$$\mathcal{P}_{1\rightarrow 2} = u_\omega B_{12}^{(\omega)} \tag{4.31}$$

and by making use of equations (4.28) and (4.26) so that

$$\frac{2\pi A^2 \omega_{12}^2}{3\hbar^2} \left| \int \psi_2^*(-e\mathbf{r})\psi_1 d\mathbf{r} \right|^2 = 2\varepsilon_0\omega_{12}^2 A^2 B_{12}^{(\omega)} \tag{4.32}$$

where we omitted the δ-Dirac term since we selected the frequency $\omega_{12} = (E_2 - E_1)/\hbar$ corresponding to the energy of the absorpted photon. After a little algebra, we get

$$B_{12}^{(\omega)} = \frac{4\pi^3}{3\varepsilon_0 h^2} \left| \int \psi_2^*(-e\mathbf{r})\psi_1 \, d\mathbf{r} \right|^2 \tag{4.33}$$

which provides the result we were in search of. By means of equation (4.13) we immediately obtain

$$B_{12}^{(\nu)} = \frac{2\pi^2}{3\varepsilon_0 h^2} \left| \int \psi_2^*(-e\mathbf{r})\psi_1 \, d\mathbf{r} \right|^2 \tag{4.34}$$

for the same coefficient in the representation of the frequency. Finally, we remark that

$$B_{12}^{(\omega)} = B_{21}^{(\omega)} \qquad B_{12}^{(\nu)} = B_{21}^{(\nu)} \qquad (4.35)$$

since the relationship between the stimulated and spontaneous absorption coefficients we already reported in equation (4.8) does not of course depend on the representation we choose.

4.2.5 Spontaneous emission coefficient

By inserting equation (4.34) into equation (4.8) it is easy to provide the microscopic expression for the spontaneous emission coefficient in the final form

$$A_{21}^{(\nu)} = \frac{16\pi^3}{3\varepsilon_0 hc^3} \nu_{21}^3 \left| \int \psi_2^*(-e\mathbf{r})\psi_1 \, d\mathbf{r} \right|^2 \qquad (4.36)$$

valid in the frequency representation.

By using the conversion provided in equation (4.13) we have

$$A_{21}^{(\omega)} = \frac{4h}{c^3} \nu_{21}^3 B_{21}^{(\omega)} \qquad (4.37)$$

so that

$$A_{21}^{(\omega)} = \frac{16\pi^3}{3\varepsilon_0 hc^3} \nu_{21}^3 \left| \int \psi_2^*(-e\mathbf{r})\psi_1 \, d\mathbf{r} \right|^2 \qquad (4.38)$$

which, interestingly enough, coincides with equation (4.36). In other words, by comparing the two expressions for $A_{21}^{(\nu)}$ and $A_{21}^{(\omega)}$ we realise that, unlike the stimulated coefficients, there is only one expression for the spontaneous coefficient, regardless of whether the representation of 'frequencies' or 'angular frequencies' is used. This result has an obvious physical meaning: the phenomenon of spontaneous emission occurs in the absence of radiation and, therefore, does not depend on how the oscillation modes of the e.m. field are represented.

The result here obtained has a direct classical counterpart. Let is consider a classic electric dipole $\mathbf{d}_{elec}$ oscillating harmonically at frequency ν. It is shown that the average power P radiated by this dipole is [2]

$$P \sim \nu^4 \, d_{elec}^2 \qquad (4.39)$$

Since dipolar radiation is quantum equivalent to the emission of photons of energy $h\nu$, we immediately obtain that the rate of emission of photons $\mathcal{R}$, i.e., the number of photons emitted per unit of time, is given by

$$\mathcal{R} = \frac{P}{h\nu} \sim \nu^3 \, d_{elec}^2 \qquad (4.40)$$

This quantity is obviously proportional to the quantum probability (per unit time) of spontaneous emission reported in equation (4.36): in fact, the two classical and quantum expressions predict identical dependence on the cube of the frequency of the emitted radiation and on the square of another mathematical object which, however, in the classical case has an immediate physical reading. In conclusion, we argue that in quantum theory the term $\int \psi_2^*(-e\mathbf{r})\psi_1 \, d\mathbf{r}$ plays the role of *an oscillating quantum dipole*.

4.3 Electric dipole selection rules for hydrogenic states

The true fact that Einstein coefficients are related to electric dipole transitions implies that *not all transitions between two arbitrarily chosen states are actually possible*. More specifically, the matrix element $\left| \int \psi_2^*(-e\mathbf{r})\psi_1 \, d\mathbf{r} \right|^2$ is certainly zero if the integrand function is odd. Since the operator $(-e\mathbf{r})$ changes sign upon inversion $\mathbf{r} \to -\mathbf{r}$, we conclude that an electric dipole transition can occur only if the initial and final states have opposite parity. This state of affairs is summarized by stating the following *electric dipole selection rule*: electric dipole transitions that would occur between two states described by wavefunctions of equal parity are generally prohibited. This statement is known as the *Laporte rule*.

In the case of hydrogenic atoms, the previous selection rule takes a more explicit form. Since the wavefunctions are

$$\psi_{nlm}(r, \theta, \phi) = R_{nl}(r) Y_{lm}(\theta, \phi) \tag{4.41}$$

it is easy to prove that: (i) the integral on the radial coordinate r always gives a result different from zero and (ii) the electric dipole matrix element is proportional to integrals like

$$\int_0^{2\pi} d\phi \int_0^\pi Y_{l_2 m_{l_2}}^*(\theta, \phi) \begin{Bmatrix} \sin\theta\cos\phi \\ \sin\theta\sin\phi \\ \cos\theta \end{Bmatrix} Y_{l_1 m_{l_1}}(\theta, \phi)\sin\theta d\theta \tag{4.42}$$

where of course the quantum numbers (l_1, m_{l_1}) and (l_2, m_{l_2}) refer to the initial and final states, respectively. The integrals given in equation (4.42) are non-zero only if

$$\left. \begin{aligned} \Delta l &= \pm 1 \\ \Delta m_l &= 0, \pm 1 \end{aligned} \right\} \text{ electric dipole selection rules} \tag{4.43}$$

once the parity of the spherical harmonic functions [3] is taken into account. We remark that this result is also valid for multi-electron atoms provided that they are treated within the central field approximation (described in the next chapter) where the spherical symmetry is preserved. In figure 4.2 some allowed emission transitions fulfilling the selection rule $\Delta l = \pm 1$ are shown in the case of the hydrogen atom.

4.4 Forbidden transitions

Transitions occurring between states for which equations (4.43) are *not* satisfied are indeed experimentally observed. They correspond to very weak spectroscopic signatures, since such transitions are much less likely than electric dipole ones.

If we reconsider the expansion given in equation (4.16) and proceed beyond the first-order, we obtain a new additive perturbation term appearing as a second-order

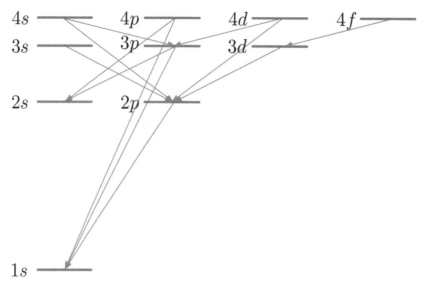

Figure 4.2. Some allowed electric dipole emission transitions fulfilling the selection rule $\Delta l = \pm 1$ in the hydrogen atom.

tensor with two components: the first one is expressed in terms of the electron orbital angular momentum, while the second one contains terms like $-exx$, $-exy$, $-exz$, ... They are, respectively, associated with *magnetic dipole and electric quadrupole transitions*. These transitions are promoted by the coupling of the magnetic field of the e.m. radiation with the orbital magnetic moment of the electron and by the coupling of the gradient of the electric field of the e.m. wave with the quadrupole electric moment of the atom.

It is proved that magnetic dipole and electric quadrupole transitions are, respectively, 10^4–10^5 and 10^7 times less probable than electric dipole ones. Their selections rules for hydrogenic atoms are calculated [3] to be

$$\left. \begin{array}{l} \Delta l = 0 \\ \Delta m_l = 0, \pm 1 \end{array} \right\} \text{ magnetic dipole selection rules} \qquad (4.44)$$

and

$$\left. \begin{array}{l} \Delta l = 0, \pm 2 \\ \Delta m_l = 0, \pm 1, \pm 2 \end{array} \right\} \text{ electric quadrupole selection rules} \qquad (4.45)$$

where in the first case the electron spin has been neglected[8]. As in the previous case, the above rules are valid in the more general case of an electron subjected to a central field.

---

[8] The complete magnetic dipole selection rules are obtained by adding the spin magnetic moment. This leads to: $\Delta l = 0$, $\Delta j = 0, \pm 1$, and $\Delta m_j = 0, \pm 1$.

4.5 The finite width of the spectral lines

Let us consider once again the case of allowed electric dipole transitions. According to equation (4.26), they should take place at exactly the same radiation frequencies as the Bohr frequencies predicted by equation (1.28). In other words, the spectral lines should be infinitely sharp. This is at odds with the experimental evidence according to which *spectral lines have a finite width*.

In order to explain this fact we will make use of the phenomenological model based on the analogy with classical physics developed at the end of the section 4.2.5. There we have associated a radiative quantum transition with a radiating classical electric dipole; this picture must be now completed by adding the notion that a charged oscillator classically undergoes radiation damping [2], because of the energy loss it experiences due to its own irradiation activity. The resulting equation of motion can be written as[9]

$$m_e \ddot{r} + 2m_e \gamma \dot{r} + m_e \omega_0^2 r = 0 \tag{4.46}$$

where it is understood that we are modelling the radiating dipole $d_{elec}(t) = -er(t)$ as a one-dimensional oscillator with (undamped) proper oscillation frequency ω_0 and mass m_e. In this equation, the term $2m_e\gamma$ is the *damping factor*. The solution of this equation is

$$r(t) = r_0 \, e^{-\gamma t} \, e^{i\omega_0 t} \tag{4.47}$$

where r_0 is the initial oscillation amplitude. This result indicates that the radiation emitted by the dumped dipole $d_{elec}(t) = -er(t)$ consists in several monochromatic Fourier components each with amplitude

$$A(\omega) = \int_0^{+\infty} r(t)e^{-i\omega t}dt = \frac{r_0}{\gamma + i(\omega_0 - \omega)} \tag{4.48}$$

so that *the intensity $A^2(\omega)$ of the radiation emitted at frequency ω* is straightforwardly calculated as

$$A^2(\omega) = \frac{r_0^2}{(\omega_0 - \omega)^2 + \gamma^2} \tag{4.49}$$

which corresponds to a Lorentzian peak, centred at ω_0 with full width at half maximum equal to 2γ. This classical result is mapped back onto the quantum picture by considering, for instance, an emission transition, as shown in figure 4.1 (left): if we identify ω_0 with the Bohr frequency $\omega_{21} = (E_2 - E_1)/\hbar$, then $(2\gamma)^{-1}$ can be interpreted as the *lifetime of the initial state E_2*.

The new fact that the initial quantum state has a finite lifetime suggests an intriguing picture, in view of the uncertainty principle discussed in section 2.3: if we look at the lifetime as the time interval Δt needed to determine E_2 of the initial state, then we draw the conclusion that this very state is actually spread over a range of energies as large as $\Delta E \sim \hbar(2\gamma)$. Accordingly, the allowed transitions do not start

---

[9] We are using the shortcuts $\ddot{r} = d^2r/dt^2$ and $\dot{r} = dr/dt$.

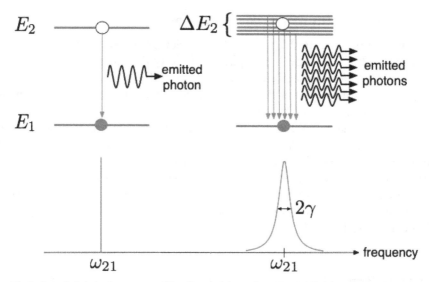

Figure 4.3. Left: an infinitely sharp spectral line (bottom) is predicted by the theory developed in section 4.2 at frequency $\omega_{21} = (E_2 - E_1)/\hbar$ for the emission transition $E_2 \to E_1$ (top). Right: the corresponding experimental situation (bottom) where the same line shows a finite width 2γ. The interpretation is that $(2\gamma)^{-1}$ is the lifetime of state E_2, reflecting the uncertainty $\Delta E_2 \sim \hbar(2\gamma)$ in the knowledge of its energy (top). For sake of clarity, similar arguments are not applied to the ground state E_1.

from an infinitely sharp energy level but, rather, from a set of spread levels, as shown in figure 4.3: this, in turn, implies that the spectral line associated with the transition shows a finite width.

Finally, we remark that the same argument developed for the initial transition state equally well applies to the final one. The most rigorous calculation of the finite width of the spectral lines does take into consideration this twofold issue: it is performed within the QED approach and eventually replaces the γ term appearing in equation (4.49) with the sum of the inverse lifetimes of the two states involved in the transition. Overall, however, QED confirms the validity of equation (4.49) and, therefore, the corresponding semi-classical picture we developed in this section.

4.6 The LASER

The normal balance between the population of the ground and the excited states dictated by Boltzmann statistics (as described in section 4.1.4) can be altered by reaching a condition such that $N_2^{(0)} > N_1^{(0)}$, usually referred to as *population inversion*. When this happens, the atomic system emits more energy than it absorbs from the electromagnetic field. This is the physical situation where the *Light Amplification by Stimulated Emission of Radiation* (LASER) phenomenon is observed.

Conceptually a LASER system is made by three elements: (i) the *active medium* where the population inversion is obtained (in principle it can be a solid, a liquid or a gas); (ii) the *energy pump*, which is any tool able to excite the active medium; (iii) the *resonating cavity*, which is typically a set of optical mirrors confining the e.m. radiation emitted by the active medium as soon as the population inversion is

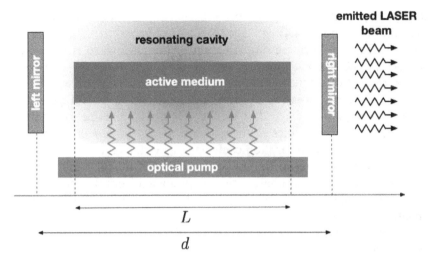

Figure 4.4. Schematic illustration of a LASER system together with its geometrical parameters d and L. In this case the device is optically pumped, that is, the pump generates a radiation field with suitable intensity and spectral characteristics tuned to excite the active medium.

generated, then favouring spontaneous emission, and finally amplifying the radiation there confined. The conceptual rendering of a LASER system is reported in figure 4.4. The light beam so generated is focussed and eventually transmitted outside the device for use as required.

4.6.1 Threshold condition for LASER emission

In order to keep the situation as simple as possible, we assume that (i) the active medium is made by a gas of atoms whose energy spectrum contains two discrete energy levels $E_1 < E_2$, and that (ii) the monochromatic e.m. radiation propagates along the z-direction, with initial intensity $I_0(\omega)$. By passing through the active medium the radiation is damped according to the Beer law [7]

$$I(\omega, z) = I_0(\omega)\exp[-\alpha(\omega)z] \qquad (4.50)$$

which contains the absorption coefficient $\alpha(\omega)$ set as

$$\alpha(\omega) = (N_1 - N_2)\sigma(\omega) \qquad (4.51)$$

where N_1 and N_2 are the populations of the two states, while $\sigma(\omega)$ is the cross-section for photon absorption: basically, it quantifies the photon capture efficiency by the gas. The population inversion is reached whenever $N_2 > N_1$ or, equivalently, $\alpha(\omega) < 0$. Because of the mirrors placed at the boundaries of the resonating cavity, the e.m. radiation is reflected back and forth, thus passing through the active medium many times. For each of such journeys, we define the *gain factor* $G(\omega)$

$$G(\omega) = \frac{I(\omega, 2L)}{I_0(\omega)} = \exp[-2\alpha(\omega)L] \qquad (4.52)$$

with $G(\omega) > 1$ provided that $\alpha(\omega) < 0$. In this equation, L indicates the width of the active medium (see figure 4.4).

In real LASER devices the amplification of the e.m. wave is counteracted by several factors, including: (i) mirrors are not ideal reflectors and, therefore, just a fraction $R < 100\%$ of the incoming light is reflected; (ii) even if the system as a whole is characterized by a negative absorption coefficient, it is still possible that a situation where $\alpha(\omega) > 0$ may occur locally, as for example due to the presence of structural defects or impurities in the active medium (moreover, mirrors are also characterized by a small absorption coefficient, as well as the material the containment system is made of); (iii) the micro-roughness of mirror surfaces generates a small component of diffuse radiation that is subtracted from the reflected radiation; (iv) after several back-and-forth cycles, the radiation undergoes diffraction phenomena (related to the curvature of the mirrors) that decrease the reflected component and, therefore, the useful light intensity. Overall their cumulative effect is summarized in the *loss factor* Γ which, for a single back-and-forth trajectory within the resonating cavity, is defined so that

$$\frac{I(\omega, 2d)}{I_0(\omega)} = \exp(-\Gamma) \tag{4.53}$$

entirely depending on the geometry and optical properties of the device (while the absorption coefficient $\alpha(\omega)$ is a characteristic of the material forming the active medium). In real devices, therefore, the gain factor is better defined as

$$G(\omega) = \frac{I(\omega, 2L)}{I_0(\omega)} \exp(-\Gamma) = \exp[-2\alpha(\omega)L - \Gamma] \tag{4.54}$$

so that the condition $G(\omega) > 1$ is reached when

$$2\alpha(\omega)L + \Gamma < 0 \tag{4.55}$$

defining the LASER as a *light amplifier* (the gain overcomes the losses). This result sharply sets the *lasing threshold condition* in terms of the population variation $\Delta N = N_2 - N_1$ as

$$\Delta N > \Delta N_{\text{threshold}} = \frac{\Gamma}{2\sigma(\omega)L} \tag{4.56}$$

meaning that, above this threshold, the LASER functioning is dictated by the stimulated emission. In other words, lasing is observed only if the optical pumping is intense enough to generate the condition $\Delta N > \Delta N_{\text{threshold}}$.

4.6.2 LASER devices

The physics of LASER devices is very complex since fundamental physics overlaps with more technological issues. In this section we outline just some qualitatively relevant aspects, referring the interested reader to specialized readings for more details [7–10].

4.6.2.1 Operational mode

There are basically *two different types of LASER devices operating in continuous and pulsed mode*, respectively.

In *continuous LASERs*, pumping is constant over time and, therefore, the sequence of physical operations can be conceptualized (with some simplifications) as follows: (i) the active medium is optically pumped and a number of electrons are excited from the ground state to the E_2 level; (ii) by spontaneous emission from this level photons are then generated; (iii) these photons begin to travel back-and-forth in the resonating cavity and, if the device is operating with a gain factor $G(\omega) > 1$, the radiation is amplified; (iv) as the number of photons trapped in the resonating cavity increases, the stimulated emission increases and, therefore, the level E_2 initially populated by the pumping is increasingly depleted; (v) a stationary condition is reached when the population (due to the constantly active pumping) and the depletion (due to the stimulated emission) phenomena are balanced; (vi) in such stationary conditions the device is able to emit (part of) the radiation generated and amplified within the resonating cavity.

In *pulsed devices*, the intensity of pumping varies over time so that the power transmitted to the active medium is described by a time pulse of duration ΔT_{pulse}. In this case, the sequence of physical operations can be conceptualized (again with some simplifications) as follows: (i) after an appropriate pumping time, at the end of which the population inversion condition has been reached, the stimulated emission is triggered; (ii) at this stage, although electrons decay from the E_2 level, the pumping pulse is still in its increasing phase and efficiently provides the level repopulation; (iii) next, as soon as the pumping pulse starts to decrease, a condition is reached in which the number of electrons leaving from the E_2 level is increasingly larger than the number of electrons there excited; (iv) ultimately, the $\Delta N < \Delta N_{\text{threshold}}$ condition is reached, stopping LASER emission. Overall, the duration of the LASER pulse is shorter than ΔT_{pulse}. There is a kind of pulsed LASERs for which the depletion process of the E_2 level is very fast. It happens, therefore, that the device soon reaches the $\Delta N < \Delta N_{\text{threshold}}$ condition even under increasing pumping and, consequently, the LASER emission terminates anyway. As the emission stops, so does the depletion: the excited level returns to being repopulated by pumping (which is nonetheless active), until the threshold condition is reached again. At this point the process is repeated iteratively several times, and the power spectrum emitted by the device consists of a sequence of very short pulses.

4.6.2.2 LASER beam alignment

Amplification is greatest for those photons that follow the longest trajectory within the active medium. Thus, the most effectively amplified photons are those travelling parallel to the axis of the resonating cavity (red photon trajectory in figure 4.5). In contrast, photons that do not travel parallel to that axis end up sooner or later intercepting the active medium, because of the multiple reflections at the mirrors. Therefore, they eventually exit the resonant condition by losing amplification (green photon trajectory in figure 4.5).

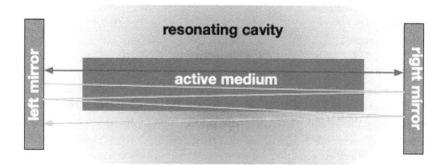

Figure 4.5. Magenta line: photon trajectory parallel to the axis of the resonating cavity and, therefore, maximally amplified. Green line: off-axis photon trajectory that eventually no longer intercepts the active medium and, therefore, loses amplification.

In conclusion, the most effectively self-sustained photon oscillations are those ones occurring around the axis of the resonating cavity with the least angular dispersion. This fact also implies that the emitted LASER beam has a reduced angular divergence and, therefore, it appears as a highly focused beam: indeed a result of very great application importance.

4.6.2.3 LASER types

There are many different types of LASER devices that differ (i) in the physical principles used for stimulated emission, (ii) the operational mode, and (iii) the spectral or power characteristics. Therefore, it is not possible to provide a universal directory; rather, one can classify LASERs according to multiple criteria and admit that each device may fall into different categories.

The first way to distinguish among different devices, is to consider the underlying *pumping principle*. In figure 4.4 and following discussion we already introduced the *optical pumping* device. On the other hand, there exist devices based on *electron pumping*: in semiconductor LASERs, for instance, the active medium is a p–n junction [11, 12] under forward bias where electrons (from the n-region) and holes (from the p-region) are injected, undergoing radiative recombination. Alternatively, a gaseous He–Ne mixture is used as active medium where an *electric discharge* is used to excite He atoms which, through interatomic collisions, release energy to Ne atoms driving them to the inversion population condition. Alternatively, Ar-based devices have been largely produced.

A second classification criterion we can adopt is based on the *emission frequency*. In *fixed frequency* LASER devices the stimulated emission occurs between two discrete energy levels and photons are emitted just at the corresponding transition frequency[10]. We remark, however, that in practice nearly all LASER devices actually work with a three-level scheme: electrons are typically pumped from a

---

[10] In the case where more than two discrete levels are used, emission is observed at any transition frequency allowed by the adopted energy level diagram.

'ground state' to high-energy intermediate (unstable) states from where they fall into the 'excited state' through non-radiative transitions (for instance, by producing waste heat). The point is that the intermediate states can be more easily occupied by excess electrons and, therefore, they act as a sort of reservoir to populate the 'excited state' for the following process of stimulated emission. The situation is conceptualized in figure 4.6. Another option is offered by the *tunable frequency* device, for instance represented by the free electron LASER (FEL) which, by a proper tuning of the operational mode, can in fact emit over a wide frequency range. Basically, an FEL consists in a beam of relativistic electrons (that is, electrons moving at a speed close to the speed of light) which, under the action of a spatially oscillating magnetic field, follow a wavy trajectory. Because of this accelerated motion, electrons emit e. m. radiation. By a suitable tuning of the spatial variation of the magnetic field, it is possible to make the radiation pulses emitted at each curved trajectory section overlap in phase, generating an extremely intense pulse. The emission frequency depends on the electron speed: by using modern particle accelerators an FEL can be tuned over a wide frequency range, varying from infrared radiation to x-rays.

Finally, we can classify LASERs according to the nature of the active medium. In *solid-state devices* it is made by a crystalline material containing lattice defects or dopant species [13] which are characterized by a set of localized discrete energy levels, providing basically the same operating scheme discussed above. In *liquid-state devices*, instead, dye molecules with a set of discrete electronic levels are diluted in a liquid and excited by any pumping mechanism. Collisions with the molecules of the liquid bring the dye molecules to the lowest roto-vibrational state of the excited electronic level[11]. From here the dye molecule undergoes radiative transitions to roto-vibrational states of the underlying electronic level. Finally, in *gas LASERs* a

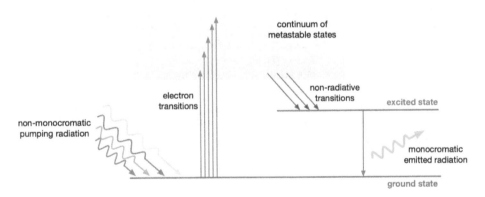

Figure 4.6. Conceptual rendering of a three-level LASER. More specifically, this picture schematically illustrates the energy diagram of a ruby LASER (see text).

---

[11] Details on the molecular roto-vibrational spectrum and electronic structure are found in chapters 7 and 8, respectively.

gaseous mixture is excited by electric discharge; the lasing activity is similar to what previously discussed in the specific case of the He–Ne LASER.

The first LASER experimentally used was the ruby LASER[12]. It is a continuous, solid-state, fixed frequency device, working under optical pumping. Figure 4.6 illustrates its three-level energy scheme. The active medium is a sapphire Al_2O_3 crystal doped by Cr^{3+} ions[13] which is illuminated by a multi-frequency flash lamp. The Cr^{3+} ions generate a continuum of energy levels, suitably placed within the energy diagram of the host material. This finite-width 'mini-band' of doping levels is able to hold all the electrons excited from the ground state by the non-mono-chromatic radiation[14]. Once excited into the 'mini-band' continuum, electrons quickly relax (typical transition times are of the order of 10^{-10} s) down to the excited level by non-radiative transitions[15]. From here, they eventually further relax to the ground state, by emitting photons just at one frequency which, in this case, corresponds to a e.m. radiation with 0.694 μm wavelength.

Further reading and references

[1] Sakurai J J 1967 *Advanced Quantum Mechanics* (Reading, MA: Addison-Wesley)

[2] Jackson J D 1975 *Classical Electromagnetism* (New York: Wiley)

[3] Bransden B H and Joachain C J 1983 *Physics of Atoms and Uolecules* (Harlow: Addison-Wesley Longman)

[4] Colombo L 2022 *Statistical Physics of Condensed Matter Systems Physics: A Primer* (Bristol: IOP Publishing)

[5] Sakurai J J and Napolitano J 2011 *Modern Quantum Mechanics* 2nd edn (Reading, MA: Addison-Wesley)

[6] Bransden B H and Joachain C J 2000 *Quantum Mechanics* (Englewood Cliffs, NJ: Prentice-Hall)

[7] Fox M 2001 *Optical Properties of Solids* (Oxford: Oxford University Press)

[8] Renk K F 2017 *Basics of Laser Physics for Students of Science and Engineering* (Heidelberg: Springer)

[9] Demtröder W 2014 *Laser Spectroscopy* vol 1 (Heidelberg: Springer)

[10] Svelto O 2010 *Principles of Lasers* (Heidelberg: Springer)

[11] Eisberg R and Resnick R 1985 *Quantum Physics of Atoms, Molecules, Solids, Nuclei, and Particles* 2nd edn (Hoboken, NJ: Wiley)

[12] Hook J R and Hall H E 2010 *Solid State Physics* (Hoboken, NJ: Wiley)

[13] Colombo L 2021 *Solid State Physics: A Primer* (Bristol: IOP Publishing)

[14] Maiman T H 1960 Stimulated optical radiation in ruby *Nature* **187** 493–4

---

[12] It is interesting to read the seminal paper by T H Maiman published in *Nature* [14].

[13] Doping basically consists in altering the pristine chemistry of the host material by insertion of other chemical species [13].

[14] In order to maximize the efficiency, the device is tailored so as to overlap the spectral region of maximum emission by the lamp with the energy interval covered by the 'mini-band' generated by the doping.

[15] This is possible, for instance, by emission of phonons, that is, quanta of the ionic vibrational field [13]. To make it simple: electrons relax by dissipating energy in form of waste heat: the temperature of the active medium is raised up by these processes.

IOP Publishing

Atomic and Molecular Physics (Second Edition)
A primer
Luciano Colombo

Chapter 5

Multi-electron atoms

Syllabus—*The basic notions developed in chapter 3 will be here extended to multi-electron atoms. Before addressing the systematic treatment of their electronic structure, we will cope with two special cases, namely the alkali atoms and the helium atom, for which a phenomenological analysis can be developed that is very useful for pedagogical purposes. In particular, the case study represented by the helium atom allows for fully exploiting the quantum mechanics of a system of indistinguishable fermions, including also the role of spin. Next, the electronic structure of non-hydrogenic atoms will be discussed under the central-field approximation, leading to a very clean and simple rationale for the construction of the periodic table of elements: the most informative summary of atomic properties. A further step will be undertaken by outlining the more advanced Hartree, Hartree–Fock, and configuration interaction theories, which describe the quantum many-body nature of the multi-electron physics with increasing accuracy. Eventually, some features regarding the excited states in multi-electron atoms, selection rules for electric dipole transitions and the action of an external magnetic field will be presented.*

5.1 Two introductory steps

5.1.1 The alkali atoms

Alkali atoms (Li, Na, K, Rb, Cs and Fr) are characterized by the presence of one electron outside an inner 'core' containing all the other ones. This electron, referred to as *the optical electron*, is on average at a much greater distance from the nucleus than all the remaining ones. Compelling experimental support for this picture is provided by some important chemical evidences: (i) their first ionization potentials are much lower than those of atoms with only one less electron (which, as we shall see, are the noble gases); (ii) in the solid form they are metals (or, equivalently, they are characterized by a conducing gas of carriers [1], basically provided by the optical electrons which, being only weakly bound to their nucleus, can easily delocalize throughout the solid); (iii) they form molecules with ionic bonding by transferring their optical electron to a cation (see discussion in section 6.3.1). Furthermore, the

doi:10.1088/978-0-7503-5734-0ch5

Lyman series in the absorption spectra of alkali atoms is very similar to the case of hydrogen[1]. More specifically, the Rydberg result provided in equation (1.17) is empirically replaced by

$$\frac{1}{\lambda} = \xi - \frac{\mathcal{R}_H}{(n - \delta_{QD})^2} \tag{5.1}$$

where ξ is a constant term fixing the position of the series head, $n = 1, 2, 3,...$ is the ordinary principal quantum number, and δ_{QD} is known as the *quantum defect*: its role is to measure the shift with respect to the hydrogen case. For this reason the quantity $n^* = (n - \delta_{QD})$ is sometimes referred to as the *effective principal quantum number* (it is no longer an integer). It is empirically determined that δ_{QD} ranges between a few tenths and a few units.

Overall, we can approach the physics of the optical electron as if it feels a screened nuclear charge because of the presence of the other electrons. A good guess for such a screened charge is

$$Z_{\text{screened}}(r) = 1 + \frac{b}{r} \tag{5.2}$$

where r is the distance from the nucleus (which we assume to be point-like) and b is a suitable constant. Our physical intuition suggests that

$$\begin{cases} \lim_{r \to 0} Z_{\text{screened}}(r) &= +Z \\ \lim_{r \to +\infty} Z_{\text{screened}}(r) &= +1 \end{cases} \tag{5.3}$$

defining the asymptotic limit of the screened potential acting on the optical electron. The intriguing new feature with respect to the case of hydrogenic atoms is that *such a potential is still central, but no longer Coulomb-like*. It is worth exploring the consequences of this.

From the body of knowledge developed in section 3.1.2 we understand that the wavefunction of an electron subjected to a central potential is cast in the form

$$\Psi_{nlm_l}(\mathbf{r}) = \bar{R}_{n,l}(r) Y_{lm_l}(\theta, \phi) \tag{5.4}$$

corresponding to a product between a radial function and a spherical harmonic function. In the case of the optical electron of an alkali atom[2], however, we used the symbol $\bar{R}_{nl}(r)$ for the radial part since this function must solve a slightly different differential equation with respect to the one treated in appendix A for the hydrogen atom. More precisely, $\bar{R}_{nl}(r)$ is found by solving

$$-\frac{\hbar^2}{2m_e} \frac{1}{r^2} \frac{d}{dr}\left[r^2 \frac{d\bar{R}(r)}{dr} \right] + \left[\frac{\hbar^2}{2m_e} \frac{l(l+1)}{r^2} - \frac{e^2}{4\pi\varepsilon_0} \frac{1}{r} - \frac{e^2}{4\pi\varepsilon_0} \frac{b}{r^2} \right] \bar{R}(r) = \bar{E}\bar{R}(r) \tag{5.5}$$

---

[1] We are not considering the fine structure.
[2] We stress that in developing this model we are not considering the remaining core electrons.

which, however, can be cast in the same form as equation (A.15) provided that we introduce an *effective orbital quantum number* l^* given by

$$\frac{\hbar^2}{2m_e}l(l+1) - \frac{e^2}{4\pi\varepsilon_0}b = \frac{\hbar^2}{2m_e}l^*(l^*+1) \tag{5.6}$$

or equivalently

$$l^*(l^*+1) = \left[l(l+1) - \frac{m_e e^2 b}{2\pi\varepsilon_0\hbar^2}\right] \tag{5.7}$$

which gives the exact definition of l^*. We can therefore calculate the energy of the optical electron as

$$\bar{E}_{n^*} = -\frac{hc\mathcal{R}_H}{(n^*)^2} \tag{5.8}$$

where n^* is the effective principal quantum number defined in equation (5.1). A good agreement with the experimental absorption spectra is obtained by setting

$$n^* = n - \delta l \tag{5.9}$$

where $\delta l = (l - l^*)$ is another way to express the quantum defect introduced above. From equation (5.7) we get

$$\delta l \simeq \frac{m_e e^2}{2\pi\varepsilon_0\hbar^2}\frac{b}{2l+1} \tag{5.10}$$

where we have neglected any quadratic term in δl. By further combining equations (5.8), (5.9), and (5.10) we eventually obtain

$$\bar{E}_{nl} = -\frac{hc\mathcal{R}_H}{\left[n - \dfrac{m_e e^2}{2\pi\varepsilon_0\hbar^2}\dfrac{b}{2l+1}\right]^2} \tag{5.11}$$

namely the energy of the quantum states of the optical electron. In figure 5.1 we compare the energy-level diagram of the hydrogen (left) and lithium (right) atoms, showing a twofold difference: (i) a downward shift of all lithium energy levels, and (ii) a resolution of the degeneracy in the l quantum number. This is in fact the most interesting consequence of equation (5.11): *the energy of the optical electron no longer depends on the principal quantum number alone, but also on the orbital quantum number*. It is therefore said that the accidental degeneracy in l has been removed due to the fact that the potential, once the screening effect due to the core electrons is properly taken into account, no longer varies as $1/r$. While the l-dependence of the energy has been here obtained for the specific case of alkali atoms by means of a phenomenological model, it nevertheless anticipates what happens in any multi-electron atom.

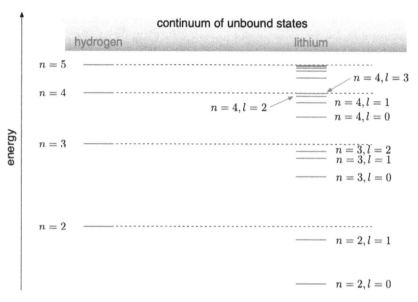

Figure 5.1. Comparison between the energy-level diagram of hydrogen (left) and lithium (right), showing the *l*-dependence in the latter case (the $n = 1$ state is not shown). No fine structure features are here considered. Figure not in scale.

5.1.2 The helium atom

5.1.2.1 Singlet–triplet states and exchange forces

In section 2.5.2 we learned that the total wavefunction describing a system of many identical and indistinguishable fermions is antisymmetric. On the other hand, we also learned that electrons carry an intrinsic angular momentum and, therefore, their single-particle wavefunction must be written as in equation (3.56). Let us consider the consequences of these two features by focussing on the specific case of *a system of two non-interacting electrons*: the corresponding total wavefunction is the product between the *total space wavefunction* Ψ (describing the orbital motion of the two independent particles) and the *total spin wavefunction* χ (describing the spin states)

$$\text{(total wavefunction)} = \text{(total space wavefuction)} \times \text{(total spin wavefunction)} \quad (5.12)$$

In order to get the proper global antisymmetry, we can combine the space and spin parts in two different ways, namely

$$\text{(total wavefunction)} = \begin{cases} \Psi_A \times \chi_S \\ \Psi_S \times \chi_A \end{cases} \quad (5.13)$$

where the labels A and S indicate, respectively, the antisymmetric or symmetric character of the function. In the specific case of two independent electrons, there is *just one possible antisymmetric total spin wavefuction*

$$\chi_A = \frac{1}{\sqrt{2}}\left[\chi_\uparrow(1)\chi_\downarrow(2) - \chi_\downarrow(1)\chi_\uparrow(2)\right] \tag{5.14}$$

and *three different symmetric total spin wavefunctions*

$$\chi_S = \begin{cases} \chi_S^{+1} = \chi_\uparrow(1)\chi_\uparrow(2) \\ \chi_S^0 = \frac{1}{\sqrt{2}}\left[\chi_\uparrow(1)\chi_\downarrow(2) + \chi_\downarrow(1)\chi_\uparrow(2)\right] \\ \chi_S^{-1} = \chi_\downarrow(1)\chi_\downarrow(2) \end{cases} \tag{5.15}$$

where the labels '1' and '2' in parenthesis indicate the two electrons, while the subscripts $\uparrow$ and $\downarrow$ indicate the spin state corresponding to the spin quantum numbers $\pm 1/2$, respectively. The superscripts -1, 0, and -1 are explained as follows. The two total spin functions describe different situations: in the state χ_A the two spins are anti-parallel, while in the states χ_S they are parallel. This is easily understood by using the vector model: the total spin $\mathbf{S}_{tot} = \mathbf{S}_1 + \mathbf{S}_2$ is quantized according to

$$S_{tot} = \sqrt{s_{tot}(s_{tot} + 1)}\,\hbar \tag{5.16}$$

$$S_{tot,z} = m_{s_{tot}}\hbar \tag{5.17}$$

where the total spin quantum number is $s_{tot} = 0, 1$. The first value corresponds to anti-parallel electron spins, while the value $s_{tot} = 1$ is found when the two spins are parallel and it is compatible with three different orientations of the total spin momentum (described by the spin magnetic numbers $m_{s_{tot}} = 0, \pm 1$). It is customary to summarize the situation here described by stating that *an antisymmetric total spin function describes a singlet state, while a symmetric one describes a triplet state.* The electron spins are anti-parallel (parallel) in the singlet (triplet) state. This justifies the notation used in equation (5.15).

The space part of the total wavefunction behaves accordingly to the singlet or triplet character of the spin part. We will develop this concept by a qualitative argument. Let us suppose that the two independent electrons can occupy two different states that we label by a compact notation as α and β, respectively[3]. We further assume that the two electrons are quite close (i.e., their coordinates have almost the same value) so that their wavefunctions are very similar in whatever state: $\psi_\alpha(1) \sim \psi_\alpha(2)$ and $\psi_\beta(1) \sim \psi_\beta(2)$. For a singlet state the total space wavefunction must be symmetric

$$\Psi_S = \frac{1}{\sqrt{2}}\left[\psi_\alpha(1)\psi_\beta(2) + \psi_\beta(1)\psi_\alpha(2)\right] \sim \sqrt{2}\psi_\alpha(1)\psi_\beta(2) \tag{5.18}$$

while for the triplet states it must be antisymmetric

---

[3] Since this discussion is only qualitative we do not need to make all quantum numbers appear explicitly: therefore, both α and β are intended as a full set of principal, angular, and magnetic quantum numbers.

$$\Psi_A = \frac{1}{\sqrt{2}}\Big[\psi_\alpha(1)\psi_\beta(2) - \psi_\beta(1)\psi_\alpha(2)\Big] \sim 0 \qquad (5.19)$$

This qualitative estimation leads to an unprecedented conclusion: *in the singlet state the two electrons have high probability to be close,* while *in the triplet states there is no way to find them nearby each other.* Phenomenologically, this situation can be described by saying that in the triplet states the two electrons repel, while in the singlet state they attract each other: *everything goes as if electrons move under the action of an 'exchange force' coupling the space and spin variables.* The exchange force has no classical counterpart and must duly be taken into account when the two electrons belong to the same atomic system.

5.1.2.2 The He ground state

The simplest of multi-electron atoms is *helium*, consisting of a nucleus with mass number $A = 4$ (two protons and two neutrons) and atomic number $Z = 2$, around which two electrons move. In order to set up the corresponding quantum problem, we will proceed step by step, similarly to the case of the hydrogen atom. First, we will assume infinite nuclear mass; next, we will neglect both magnetic interactions and relativistic effects. The remaining interactions will be, therefore, only of electrostatic nature. Once again, the frame of reference will be centred on the nucleus, as shown in figure 5.2.

Under the above assumptions the energy eigenvalue equation is written as

$$\left[-\frac{\hbar^2}{2m_e}(\nabla_1^2 + \nabla_2^2) - \frac{1}{4\pi\varepsilon_0}\frac{Ze^2}{r_1} - \frac{1}{4\pi\varepsilon_0}\frac{Ze^2}{r_2} + \frac{1}{4\pi\varepsilon_0}\frac{e^2}{r_{12}}\right]\Psi(\mathbf{r}_1, \mathbf{r}_2) = E\,\Psi(\mathbf{r}_1, \mathbf{r}_2) \quad (5.20)$$

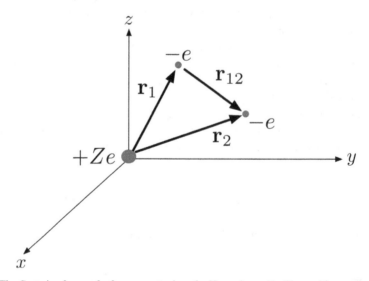

Figure 5.2. The Cartesian frame of reference centred on the He nucleus $+Ze$. The position vectors $\mathbf{r}_1$ and $\mathbf{r}_2$ of the two electrons $-e$ are shown, together with their relative distance $\mathbf{r}_{12}$.

where ∇_i^2 is calculated with respect to the coordinates of the ith electron ($i = 1, 2$) and $\Psi(\mathbf{r}_1, \mathbf{r}_2)$ is the total space wavefunction for the two-electron system. The first term in square parenthesis represents the total electron kinetic energy, the next two terms represent the electron–nucleus Coulomb interactions, while the last term describes the electron–electron coupling[4].

In order to proceed we introduce the so-called *independent electron approximation* (or *single-electron picture*): simply, *the electron–electron interaction is in a first instance neglected* (and it will be considered afterwards as a perturbation to such a zeroth-order approximation). Under this assumption equation (5.20) is separated into two independent single-electron problems and, therefore, the total wavefunction $\Psi(\mathbf{r}_1, \mathbf{r}_2)$ is just the product of two hydrogenic wavefunctions $\psi^{(h)}$, where the superscript '(h)' indicates that we have applied all the hydrogenic corrections described in section 3.2. Therefore, in the zeroth-order approximation (hereafter labelled by the superscript '(0)') for the ground state (GS) of He atom we can write

$$\Psi_{GS}^{(0)}(\mathbf{r}_1, \mathbf{r}_2) = \psi_{1s}^{(h)}(\mathbf{r}_1)\psi_{1s}^{(h)}(\mathbf{r}_2) \quad E_{GS}^{(0)} = 2Z^2 E_{1s}^{(H)} \simeq -108.8 \text{ eV} \tag{5.21}$$

where the superscript '(H)' now indicates the corresponding quantity of the true hydrogen atom. The experimental value for the ground state energy $E_{GS}^{expt} = -78.62$ eV is quite far from the above prediction, indicating that the electron–electron coupling must be necessarily added to the zeroth-order picture. Since the corresponding perturbation operator $\hat{H}_{pert} = e^2/4\pi\varepsilon_0 r_{12}$ does not depend on time, we can use equation (2.96) and write

$$E_{GS} = E_{GS}^{(0)} + \int \psi_{1s}^{(h)*}(\mathbf{r}_1)\psi_{1s}^{(h)*}(\mathbf{r}_2) \frac{e^2}{4\pi\varepsilon_0 r_{12}} \psi_{1s}^{(h)}(\mathbf{r}_1)\psi_{1s}^{(h)}(\mathbf{r}_2) \, d\mathbf{r}_1 d\mathbf{r}_2 \tag{5.22}$$

$$= E_{GS}^{(0)} + \underbrace{\frac{1}{4\pi\varepsilon_0} \int \frac{(-e)|\psi_{1s}^{(h)}(\mathbf{r}_1)|^2 \, (-e)|\psi_{1s}^{(h)}(\mathbf{r}_2)|^2}{r_{12}} \, d\mathbf{r}_1 d\mathbf{r}_2}_{= I_{1s,1s}} \tag{5.23}$$

where we have defined the *Coulomb integral* $I_{1s,1s}$ representing the classical electrostatic energy of two interacting charge densities $\rho(\mathbf{r}_1) = (-e)|\psi_{1s}^{(h)}(\mathbf{r}_1)|^2$ and $\rho(\mathbf{r}_2) = (-e)|\psi_{1s}^{(h)}(\mathbf{r}_2)|^2$ placed at distance r_{12}. The explicit calculation [2–4] leads to

$$I_{1s,1s} = -\frac{5}{4} Z E_{1s}^{(H)} \tag{5.24}$$

so that to the first-order in the perturbation we get

$$E_{GS} = E_{GS}^{(0)} + I_{1s,1s} = \left(2Z^2 - \frac{5}{4}Z\right)E_{1s}^{(H)} \simeq -74.8 \text{ eV} \tag{5.25}$$

in much better agreement with the experimental data.

---

[4] More rigorously, we should say that each term represents the quantum operator corresponding to any single energy contribution. For sake of simplicity, hereafter we will omit this mathematical subtlety.

A better result is found by following a different approach, namely by considering the *screening effects* (still within the single-electron picture): because of the presence of the other electron, each electron does not actually feel a bare nuclear charge but, rather, a screened one. It is possibile to demonstrate that the screened nuclear charge for the He ground state is $Z^* = Z - 5/16$ providing a corrected ground state energy

$$E_{GS} = \left(2Z^2 - \frac{5}{4}Z + \frac{25}{128}\right)E_{1s}^{(H)} \simeq -77.45 \text{ eV} \tag{5.26}$$

This calculation is reported in appendix G, while the separate effect of (i) the electron–electron coupling, and (ii) the nuclear screening is shown in figure 5.3 with respect to the crudest independent electron model.

5.1.2.3 The He excited states and the exchange interactions

The physics of the excited states in the He atom can be described by following the same approach as above: a zeroth-order single-particle picture is first elaborated and next corrected by evaluating the electron–electron coupling as a perturbation. In this case, the first electron is placed in the $1s$ hydrogenic state, while the second one occupies the state described by the quantum numbers nl with $n > 1$ (for sake of simplicity we neglect screening effects). It is therefore straightforward to write

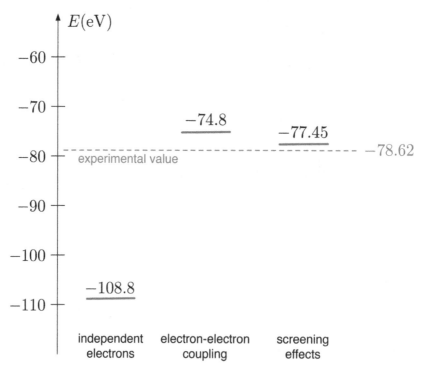

Figure 5.3. The energy of the ground state of the He atom calculated within the crude single-particle picture (left) and, next, by including the electron–electron coupling (middle) or by correcting with screening effects (right).

$$\Psi^{(0)}_{1s,\,nl}(\mathbf{r}_1,\,\mathbf{r}_2) = \psi^{(h)}_{1s}(\mathbf{r}_1)\psi^{(h)}_{nl}(\mathbf{r}_2) \quad E_{1s,nl} = Z^2 E^{(H)}_{1s} + Z^2 E^{(H)}_{nl} + I_{1s,nl} \tag{5.27}$$

where we have introduced the Coulomb integral

$$I_{1s,nl} = \frac{e^2}{4\pi\varepsilon_0} \int \left| \psi^{(h)}_{1s}(\mathbf{r}_1) \right|^2 \frac{1}{r_{12}} \left| \psi^{(h)}_{nl}(\mathbf{r}_2) \right|^2 d\mathbf{r}_1 d\mathbf{r}_2 \tag{5.28}$$

which generates the energy spectrum reported in figure 5.4. It is interesting to observe that while the ground state is non-degenerate, the electron–electron coupling splits excited states according to the specific value of the Coulomb integrals. Their calculation is very cumbersome [2–4] and, therefore, we limit to provide some order-of-magnitude estimation: $I_{1s,2s} \simeq 9$ eV and $I_{1s,2p} \simeq 10$ eV.

If we compare the best prediction we elaborated so far (corresponding to the middle spectrum of figure 5.4) with the experimental information deduced by spectroscopic measurements (corresponding to the right spectrum of figure 5.4) we immediately realise that, apart from the ground state, *the experimental spectrum is richer since it shows a double sequence of levels*. This clearly indicates that something important is still missing in our theoretical picture: basically, *we omitted to include the exchange forces discussed in section 5.1.2*. This can be done still not including explicitly the spin in our model, but fully exploiting the antisymmetric or symmetric character of the space wavefunction $\Psi(\mathbf{r}_1,\,\mathbf{r}_2)$ which has not been taken into account so far.

Let us treat exchange interactions as a perturbation on the zeroth-order description of the excited states provided by equation (5.27). The unperturbed wavefunctions for the singlet and triplet states are written as

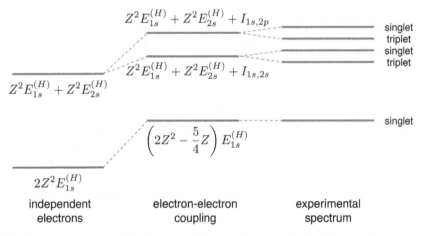

Figure 5.4. The energy spectrum of the He atom. Left: energy levels obtained within the independent electron approximation. Middle: energy levels including electron–electron coupling. Right: the experimental spectrum. The left and middle spectra are obtained neglecting screening effects. The picture is not in scale.

$$\Psi_S(\mathbf{r}_1, \mathbf{r}_2) = \frac{1}{\sqrt{2}}\left[\psi_{1s}^{(h)}(\mathbf{r}_1)\psi_{nl}^{(h)}(\mathbf{r}_2) + \psi_{1s}^{(h)}(\mathbf{r}_2)\psi_{nl}^{(h)}(\mathbf{r}_1)\right] \text{ singlet state}$$

$$\Psi_A(\mathbf{r}_1, \mathbf{r}_2) = \frac{1}{\sqrt{2}}\left[\psi_{1s}^{(h)}(\mathbf{r}_1)\psi_{nl}^{(h)}(\mathbf{r}_2) - \psi_{1s}^{(h)}(\mathbf{r}_2)\psi_{nl}^{(h)}(\mathbf{r}_1)\right] \text{ triplet states}$$

(5.29)

so that the corrected energy for singlet and triplet excited state is calculated as

$$E_{1s,\,nl}^{\text{singlet}} = Z^2 E_{1s}^{(H)} + Z^2 E_{nl}^{(H)} + I_{1s,nl} + K_{1s,nl}$$
$$E_{1s,\,nl}^{\text{triplet}} = Z^2 E_{1s}^{(H)} + Z^2 E_{nl}^{(H)} + I_{1s,nl} - K_{1s,nl}$$

(5.30)

where we have introduced the *exchange integral*[5]

$$K_{1s,nl} = \int \psi_{1s}^{(h)*}(\mathbf{r}_1)\psi_{nl}^{(h)*}(\mathbf{r}_2)\,\frac{e^2}{4\pi\varepsilon_0 r_{12}}\,\psi_{1s}^{(h)}(\mathbf{r}_2)\psi_{nl}^{(h)}(\mathbf{r}_1)\,d\mathbf{r}_1 d\mathbf{r}_2 \;>\; 0$$

(5.31)

whose value is much smaller than that of the Coulomb integrals; for instance it is calculated $K_{1s,2s} \simeq 0.4\,\text{eV}$ and $K_{1s,2p} \simeq 0.1\,\text{eV}$ [4]. The physical origin of such a positive energy term is purely quantum mechanical, with no analog in classical physics: it is due to the dynamical equivalence of the two indistinguishable electrons according to which it is conceptually wrong to set the total wavefunction in the form given in equation (5.27) since we are not allowed to select the electron '1' and place it on the hydrogenic quantum state $1s$ and, similarly, to select the electron '2' and place it on the nl state. We must rather set the total wavefunction as in equation (5.29) which, presenting a twofold option, naturally leads to the existence of a double sequence of levels, corresponding to singlet and triplet states. It is customary to say that *singlet quantum states describe the energy spectrum of para-helium*, while *triplet quantum states describe the energy spectrum of ortho-helium*. Since $K_{1s,nl} > 0$ [2, 3], then equation (5.30) proves that for any given set nl of quantum numbers *ortho-helium states have lower energy than the para-helium ones*, the energy splitting being easily calculated as $\Delta E_{\text{singlet-triplet}} = 2K_{1s,nl}$. The effect of exchange interactions is shown in figure 5.5 for the helium excited state with $n = 2$. The picture we elaborated is also able to explain another experimental feature, namely the singlet character of the ground state.

5.1.2.4 The total wavefunction and the selection rules

In order to provide the most complete description of the total wavefunction, we must add its spin part according to the discussion developed in section 5.1.2. *Para*-helium singlet states will be described by a sole wavefunction

$$\Phi^{\text{para}}(\mathbf{r}_1, \mathbf{r}_2) = \Psi_S(\mathbf{r}_1, \mathbf{r}_2)\,\chi_A$$

(5.32)

where the spin function χ_A is given in equation (5.14), while *ortho*-helium triplet states are described by the following set of wavefunctions

---

[5] The explicit calculation of an exchange integral is rather complicated: the interested reader can find it elsewhere [4].

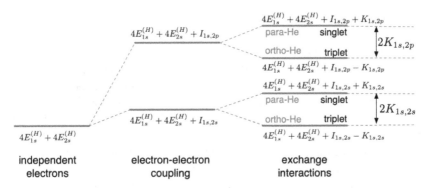

Figure 5.5. The energy of the $n = 2$ excited state of the He atom calculated within the single-particle picture (left), including Coulomb interactions (middle), and further corrected by adding exchange interactions (right). The energies E_1 and E_2 correspond to the ground state and first excited state energies of the hydrogen atom, respectively; $I_{1s,nl}$ are the Coulomb integrals; $K_{1s,nl}$ are the exchange integrals. The picture is not in scale.

$$\Phi^{\mathrm{ortho}}(\mathbf{r}_1, \mathbf{r}_2) = \begin{cases} \Psi_A(\mathbf{r}_1, \mathbf{r}_2)\,\chi_S^{+1} \\ \Psi_A(\mathbf{r}_1, \mathbf{r}_2)\,\chi_S^{0} \\ \Psi_A(\mathbf{r}_1, \mathbf{r}_2)\,\chi_S^{-1} \end{cases} \tag{5.33}$$

where the spin functions χ_S are given in equation (5.15).

Electric dipole transitions are experimentally observed provided that the initial and final states have the same spinor character. In other words, spectroscopic measurements provide evidence that *transitions between singlet and triplet states are forbidden*. This feature is explained by extending the arguments reported in section 4.3 to the case of wavefunctions described in equations (5.32) and (5.33). We imagine any (absorption) excitation as if one electron is left in the ground state, while the second one is excited to a different hydrogenic state[6]. The total electron dipole moment of an He atom is proportional to the sum $\mathbf{r}_1 + \mathbf{r}_2$ and therefore its matrix element is calculated as

$$\int \Phi^*_{\mathrm{final}}\,[\mathbf{r}_1 + \mathbf{r}_2]\,\Phi_{\mathrm{initial}}\,d\mathbf{r}_1 d\mathbf{r}_2$$
$$= \left[\int \chi^*_{\mathrm{final}}\,\chi_{\mathrm{initial}}\,ds_1 ds_2\right]\left[\int \Psi^*_{\mathrm{final}}\,[\mathbf{r}_1 + \mathbf{r}_2]\,\Psi_{\mathrm{initial}}\,d\mathbf{r}_1 d\mathbf{r}_2\right] \tag{5.34}$$

where we have labelled by 'initial' or 'final' the wavefunctions describing the two states involved in the transitions. The first integral appearing in the right hand side of the above equation is performed on the spin coordinates s_1 and s_2 of the two electrons and immediately leads to

$$\int \chi^*_{\mathrm{final}}\,\chi_{\mathrm{initial}}\,ds_1 ds_2 = 0 \tag{5.35}$$

---

[6] Our argument is easily extended to the opposite case of emission transition.

for states with opposite spin because of the orthogonality of the corresponding spin functions. This means that it holds a *selection rule* $\Delta s = 0$ for the electron undergoing the transition: if the initial state was a singlet (triplet) one, so it will have to be even after the transition. This conclusion is further confirmed by the behaviour of the integral on the space coordinates: if the initial and final spin states were χ_A and χ_S (or vice versa), then the space functions should necessarily be Ψ_S and Ψ_A (or vice versa) and the integrand function would accordingly change sign by inversion $r_1 \leftrightarrow r_2$. This forces the integral to be zero, since its value cannot depend on the arbitrary labelling of the two electrons. Finally, the integral of the space coordinates sets a second selection rule $\Delta l = \pm 1$ for the orbital quantum number of the electron undergoing the transition. In this case the argument proceeds exactly as reported in section 4.3.

We conclude this excursion on the electric dipole selection rules holding for the He atom by observing that they have been obtained under the assumption of no relativistic effects. In particular, we neglected spin–orbit interactions. While, similar to hydrogenic atoms, they are less intense than Coulomb interactions, as a matter of fact they are not exactly zero. Therefore, the wavefunctions provided in equations (5.32) and (5.33) are not rigorously the proper wavefunctions when spin–orbit interactions are taken into account. This reflects in the fact that transitions with $\Delta s \neq 0$ are indeed observed in high-precision spectroscopic measurements, although they are very weak. For atoms with larger Z values, their intensity grows accordingly.

5.1.2.5 The Heisenberg picture

We previously reported that exchange integrals are positive: the effective repulsive interaction (i.e., the net Coulomb+exchange effect) is weaker for two electrons with parallel spins (triplet state) and this implies that they correlate so as, on average, to place at the maximum possible distance. This is consistent with the discussion reported in section 5.1.2.

According to the *Heisenberg picture*, the separation between electrons due to exchange forces can be attributed to a spin–spin interaction

$$E_{spin-spin} = -2K \, \mathbf{S}_1 \cdot \mathbf{S}_2 \tag{5.36}$$

where K is a suitable exchange integral, while $\mathbf{S}_{1,2}$ are the spins of the two electrons. In order to get the proper units for the spin–spin interaction energy, we must assume that in this equation both spins are written in units of $\hbar$. By using the vector model we set $\mathbf{S}_{tot} = \mathbf{S}_1 + \mathbf{S}_2$ and calculate

$$\begin{aligned}
\mathbf{S}_1 \cdot \mathbf{S}_2 &= \frac{1}{2}[s_{tot}(s_{tot}+1) - s_1(s_1+1) - s_2(s_2+1)] \\
&= \frac{1}{2}\left[s_{tot}(s_{tot}+1) - \frac{3}{2}\right]
\end{aligned} \tag{5.37}$$

where $s_1 = s_2 = 1/2$ are the two single-electron spin quantum numbers, while s_{tot} is the total spin quantum number. In a singlet state (anti-parallel spins) we have

$s_{tot} = 0$ and, therefore, $\mathbf{S}_1 \cdot \mathbf{S}_2 = -3/4$. On the other hand, in triplet states (parallel spins) we have $s_{tot} = 1$ and, therefore, $\mathbf{S}_1 \cdot \mathbf{S}_2 = +1/4$. This implies that the spin–spin interaction is

$$
E_{\text{spin-spin}} = \begin{cases} +\dfrac{3}{2}K & \text{singlet states} \\[2em] -\dfrac{1}{2}K & \text{triplet states} \end{cases}
\tag{5.38}
$$

which provides the same energy splitting $\Delta E_{\text{singlet-triplet}} = 2K$ calculated in section 5.1.2.

Apart from providing a different explanation for an already known result, the Heisenberg picture is useful because it states that *for each given electronic configuration, the ground state corresponds to the one with parallel electronic spins*. The generalization of this conclusion is often stated in terms of the so-called *Hund rule*: the electrons placed on the generic nl configuration tend to maximize the total spin. This rule will be very useful when looking for the ordering criteria underlying the periodic table of elements.

5.2 The central-field approximation

Any attempt to extend the perturbative approach previously adopted for He to atoms containing more than two electrons becomes more and more complicated as the number of electrons grows: a radically new strategy is needed, although it is useful to still neglect both magnetic and relativistic effects in a first approach.

Let us consider an atom containing N *spinless electrons* and let us assume its nuclear mass to be infinite. The Hamiltonian operator containing just electrostatic interactions is written as

$$
\hat{H} = -\frac{\hbar^2}{2m_e} \sum_{i=1}^{N} \nabla_i^2 - \frac{Ze^2}{4\pi\varepsilon_0} \sum_{i=1}^{N} \frac{1}{r_i} + \frac{e^2}{4\pi\varepsilon_0} \sum_{i>j=1}^{N} \frac{1}{r_{ij}}
\tag{5.39}
$$

where, similarly to the previous cases, the electron positions $\mathbf{r}_i$ (with $i = 1, 2, 3,\ldots, N$) are calculated with respect to a Cartesian frame of reference centred on the nucleus. The distance between the ith and the jth electron is indicated as r_{ij}. We remark once again that spin effects are not included in the Hamiltonian model set up in the above equation: they will be duly taken into account in a second step described below. The corresponding energy eigenvalue problem is set as

$$
\hat{H} \ \Psi(\mathbf{r}_1, \mathbf{r}_2, \mathbf{r}_3,\ldots, \mathbf{r}_N) = E_{\text{tot}} \ \Psi(\mathbf{r}_1, \mathbf{r}_2, \mathbf{r}_3,\ldots, \mathbf{r}_N)
\tag{5.40}
$$

where E_{tot} is the total energy of the N-electron system and $\Psi(\mathbf{r}_1, \mathbf{r}_2, \mathbf{r}_3,\ldots, \mathbf{r}_N)$ is the corresponding total space wavefunction.

The N-body mathematical problem defined in equation (5.40) is not separable and it has no analytical solution. Furthermore, we do not expect that completely neglecting the electron–electron interaction term (as we did for He in a first rough estimate) can work in this case: its weight is likely so large that such interactions can be neither overlooked nor even treated as a perturbation. We will rather adopt a different theoretical scheme, originally introduced by Slater and Hartree [2, 4], which is based on two pillars, namely the *independent electron picture* and the *central field approximation*. Basically, it is assumed that *each single electron moves under the action of an effective central-field potential $V_{cf}(r_i)$ ($i = 1, 2, 3,\ldots, N$) describing in average the effects of the interactions with any other electron and with the nucleus.*

While the actual expression for $V_{cf}(r_i)$ is still unknown, its very definition dictates the spherical symmetry and our physical intuition leads to guessing its behaviour when the ith electron is found either within the core atomic region or very far away from the nucleus. These situations correspond to the following limiting cases valid for a neutral atom ($Z = N$)

$$V_{cf}(r_i) = \begin{cases} -\dfrac{[Z - (N - 1)]e^2}{4\pi\varepsilon_0}\dfrac{1}{r_i} = -\dfrac{e^2}{4\pi\varepsilon_0}\dfrac{1}{r_i} & \text{when } r_j \ll r_i \to +\infty \ \ \forall j \neq i \\[4mm] -\dfrac{Ze^2}{4\pi\varepsilon_0}\dfrac{1}{r_i} + \dfrac{(N - 1)e^2}{4\pi\varepsilon_0}\dfrac{1}{r_{ave}} & \text{when } r_j \gg r_i \to 0 \ \ \forall j \neq i \end{cases}$$ (5.41)

where r_{ave} is the average orbital radius of the remaining $(N - 1)$ electrons[7]. The determination of the actual $V_{cf}(r_i)$ for any electron at intermediate distance $0 \ll r_i \ll +\infty$ will be discussed in section 5.2.1: for the moment we assume it is known, so that we can formally proceed in developing the central-field approximation.

By introducing into equation (5.39) the operator $\hat{V}_{cf}(r_i)$ corresponding to the central-field potential, we can recast the total Hamiltonian operator in the form

$$\hat{H} = -\frac{\hbar^2}{2m_e}\sum_{i=1}^{N}\nabla_i^2 - \frac{Ze^2}{4\pi\varepsilon_0}\sum_{i=1}^{N}\frac{1}{r_i} + \frac{e^2}{4\pi\varepsilon_0}\sum_{i>j=1}^{N}\frac{1}{r_{ij}}$$

$$= \underbrace{-\frac{\hbar^2}{2m_e}\sum_{i=1}^{N}\nabla_i^2 + \sum_{i=1}^{N}\hat{V}_{cf}(r_i)}_{\hat{H}_{cf}} \underbrace{-\sum_{i=1}^{N}\left[\frac{Ze^2}{4\pi\varepsilon_0}\frac{1}{r_i} + \hat{V}_{cf}(r_i)\right] + \frac{e^2}{4\pi\varepsilon_0}\sum_{i>j=1}^{N}\frac{1}{r_{ij}}}_{\hat{H}_{corr}}$$ (5.42)

$$= \qquad\qquad \hat{H}_{cf} \qquad\qquad + \qquad\qquad \hat{H}_{corr}$$

where *the complete problem is separated into a 'central-field' part described by the operator $\hat{H}_{cf}$ and a corrective part described by the operator $\hat{H}_{corr}$* containing both spherical ($\sum_i[Ze^2/4\pi\varepsilon_0 r_i + \hat{V}_{cf}(r_i)]$) and non-spherical ($\sum_{i>j}e^2/4\pi\varepsilon_0 r_{ij}$) corrections to the central field. The key point is that the interactions described by $\hat{H}_{corr}$ are small as

---

[7] We remark that the term $(N - 1)e^2/4\pi\varepsilon_0 r_{ave}$ appearing in the second equation is just a constant that can be gauged to zero with no loss of physical information.

compared to those included in $\hat{H}_{cf}$: accordingly, we will neglect them and we will correct the central-field picture by different methods, as shown below[8].

By replacing the true energy eigenvalue problem given in equation (5.40) by its central-field approximation, we can write

$$\sum_{i=1}^{N}\left[-\frac{\hbar^2}{2m_e}\nabla_i^2 + \hat{V}_{cf}(r_i)\right]\Psi(\mathbf{r}_1, \mathbf{r}_2, \mathbf{r}_3,\dots, \mathbf{r}_N) = E_{tot}\,\Psi(\mathbf{r}_1, \mathbf{r}_2, \mathbf{r}_3,\dots, \mathbf{r}_N) \qquad (5.43)$$

where, interestingly enough, each term appearing in the square parenthesis only depends on the coordinate of a single electron. This has a twofold consequence: (i) the total Hamiltonian operator is conveniently cast in the form of a sum of single-electron Hamiltonian operators

$$\hat{h}_i = -\frac{\hbar^2}{2m_e}\nabla_i^2 + \hat{V}_{cf}(r_i) \quad \rightarrow \hat{H}_{cf} = \sum_{i=1}^{H}\hat{h}_i \qquad (5.44)$$

and (ii) we can separate the variables in the total wavefunction by setting

$$\Psi(\mathbf{r}_1, \mathbf{r}_2, \mathbf{r}_3,\dots, \mathbf{r}_N) = \prod_{i=1}^{N}\psi_{\alpha_i}(\mathbf{r}_i) \qquad (5.45)$$

where α_i is a generic label to indicate the set of suitable quantum numbers describing the vector state (space part) of the ith electron. The N-body problem is replaced by N single-electron problems

$$\left[-\frac{\hbar^2}{2m_e}\nabla_i^2 + \hat{V}_{cf}(r_i)\right]\psi_{\alpha_i}(\mathbf{r}_i) = E_{\alpha_i}\,\psi_{\alpha_i}(\mathbf{r}_i) \qquad (5.46)$$

where the individual energies E_{α_i} sum up

$$E_{tot,cf} = \sum_{i=1}^{N}E_{\alpha_i} \qquad (5.47)$$

thus providing *the central-field estimation for the total energy of the electron system*.

Since $V_{cf}(r_i)$ is by construction a central potential, the corresponding central-field wavefunctions $\psi_{\alpha_i}(\mathbf{r}_i)$ are the product of a radial part $\bar{R}_{n_i l_i}(r_i)$ and a spherical harmonic function $Y_{l_i m_{l_i}}(\theta_i, \phi_i)$

$$\psi_{\alpha_i}(\mathbf{r}_i) = \bar{R}_{n_i l_i}(r_i)\,Y_{l_i m_{l_i}}(\theta_i, \phi_i) \qquad (5.48)$$

where the three quantum numbers $\{n_i, l_i, m_{l_i}\} = \alpha_i$ have the same meaning and obey the same rules as in the case of hydrogenic atoms. The most important difference is that *the radial central-field functions $\bar{R}_{n_i l_i}(r_i)$ are not those ones reported in table 3.2*

---

[8] In a more thorough approach $\hat{H}_{corr}$ can be treated as a perturbation on the eigenstates of $\hat{H}_{cf}$. This is technically demanding calculation which is necessary only when the highest degree of rigour and accuracy is indeed required.

since the central-field potential does not in general behave as $\sim 1/r$ as in the case of hydrogenic atoms. On the other hand, the Y-functions are true spherical harmonics. Another consequence of this feature is that *single-electron energies depend upon both the principal and the angular momentum quantum numbers*, since the accidental degeneracy is removed, as anticipated in section 5.1.1 by treating the specific case of alkali atoms. Accordingly, any multi-electron electronic configuration is assigned by setting all pairs (n_i, l_i) for any $i = 1, \ldots, N$.

In concluding the discussion of the central-field approximation we must observe that the total wavefunction given in equation (5.45) is invariant upon any permutation of atomic coordinates, since electrons are identical and indistinguishable particles. This feature is recognized as the *exchange degeneracy*: a feature that will be duly readdressed in section 5.2.2.

5.2.1 The self-consistent-field method

In order to proceed with the development of our quantum model we need to define, at least conceptually (the practical implementation is a rather challenging task, more suitably treated in quantum chemistry textbooks [5]) a procedure for calculating the classical expression of $V_{cf}(r_i)$. A very effective way to accomplish this goal was first devised by Herman and Skillman in 1960, and has proved to be particularly suitable for numerical implementations and, therefore, it is nowadays routinely adopted (at very different levels of sophistication) in modern computer-based electronic-structure calculations.

Let us start with a set $\{\psi_{\alpha_j}(\mathbf{r}_j)\}$ of trial functions[9] to define a zeroth-order form of the self-consistent potential acting on the ith electron

$$V_{cf}^{(0)}(r_i) = -\frac{1}{4\pi\varepsilon_0}\frac{Ze^2}{r_i} + \frac{e^2}{4\pi\varepsilon_0}\sum_{j>i=1}^{N}\int\frac{\left|\psi_{\alpha_j}(r_j)\right|^2}{r_{ij}}\,d\mathbf{r}_j \qquad (5.49)$$

This potential form is inserted into equation (5.46), the equation is solved, and a new set $\{\phi_{\beta_j}(\mathbf{r}_j)\}$ of eigenfunctions is calculated. A check is now operated to verify whether the ϕ-functions are equal to the initial ψ-ones. If not (as is the case for the first iteration), a new first-order corrected form of the self-consistent potential is accordingly obtained as

$$V_{cf}^{(1)}(r_i) = -\frac{1}{4\pi\varepsilon_0}\frac{Ze^2}{r_i} + \frac{e^2}{4\pi\varepsilon_0}\sum_{j>i=1}^{N}\int\frac{\left|\phi_{\beta_j}(r_j)\right|^2}{r_{ij}}\,d\mathbf{r}_j \qquad (5.50)$$

and the same calculation as above is again executed, producing a third set of γ-functions. This procedure is iteratively repeated, generating at each next iteration a new set of eigenfunctions of increasing accuracy. The iteration proceeds until it is

---

[9] A possible choice is taking hydrogenic functions, or some linear combination of them, or Gaussian-like functions. Alternatively, wide use of the so-called Slater-type orbitals has been made [6, 7].

obtained that $(n-1)$th-order wavefunctions are equal (within an agreed degree of numerical accuracy) to the nth-order ones. At this point the 'full converge' is proclaimed and the calculation stopped. This procedure allows one to accurately determine the radial part $\bar{R}_{n_i l_i}(r_i)$ of the electron wavefunctions as well as their energies $E_{n_i l_i}$.

An alternative protocol consists in guessing a zeroth-order spherical effective potential of the form

$$V_{\text{cf}}^{(0)}(r_i) = -\frac{1}{4\pi\varepsilon_0} \frac{Z(r_i)e^2}{r_i} \tag{5.51}$$

and then following an iterative calculation as above. The guidelines for setting $V_{\text{cf}}^{(0)}(r_i)$ are (i) $\lim_{r_i \to 0} Z(r_i) = Z$ and (ii) $\lim_{r_i \to \infty} Z(r_i) = 1$.

Finally, it must be remembered that the determination of the central field potential $V_{\text{cf}}(r)$ was historically obtained within the so-called *Thomas–Fermi model* which is briefly presented in appendix H. The corresponding Thomas–Fermi potential can be used for many purposes, including considering it as the first guess for a self-consistent procedure.

5.2.2 Including spin-related features

The central-field approximation has been so far developed by treating electrons as spinless particles. We can overcome this approximation by simply multiplying the wavefunctions given in equation (5.48) by the spin function provided in equation (3.57), thus defining for each ith electron its *central-field spin-orbital*

$$\psi_{\alpha_i}(\mathbf{r}_i) = \bar{R}_{n_i l_i}(r_i) \, Y_{l_i m_{l_i}}(\theta_i, \phi_i) \, \chi_{s_i m_{s_i}} \tag{5.52}$$

which are characterized by the fourfold set of quantum numbers $\{n_i, l_i, m_{l_i}, m_{s_i}\}$. This is indeed an important step forward: we can now straightforwardly correct the fact that the total wavefunction $\Psi(\mathbf{r}_1, \mathbf{r}_2, \mathbf{r}_3, \ldots, \mathbf{r}_N)$ provided in equation (5.45) does not fulfil the Pauli principle (see section 2.5.2) since it is not antisymmetric. Simply, the *true central-field total wavefunction* $\Psi_{\text{cf}}(1, 2, \ldots, N)$ is written as a Slater determinant

$$\Psi_{\text{cf}}(1, 2, \ldots, N) = \frac{1}{\sqrt{N!}} \begin{vmatrix} \psi_\alpha(1) & \psi_\alpha(2) & \cdots & \psi_\alpha(N) \\ \psi_\beta(1) & \psi_\beta(2) & \cdots & \psi_\beta(N) \\ \cdots & \cdots & \cdots & \cdots \\ \psi_\nu(1) & \psi_\nu(2) & \cdots & \psi_\nu(N) \end{vmatrix} \tag{5.53}$$

where for simplicity we used the compact notation $\mathbf{r}_i \to i$ and $\{n, l, m_l, m_s\} \to a$ Greek letter.

The final result in the above determinant form makes it clear that, although we are operating within an independent particle picture, it is not possible for two electrons to have the same set of four quantum numbers: if they lie on the same energy level E_{nl} they must differ at least for their spin orientation. This will be the *guideline for building up the ground-state electronic configuration of any multi-electron atom.*

5.3 The periodic system of the elements

One of the most important results of the central-field approximation is to allow the *determination of the electronic configuration for the ground-state of any multi-electron atom* through a conceptually very simple procedure. We just need to implement two criteria: (i) the total energy given in equation (5.47) must be the absolute minimum one; and (ii) the electrons must obey the Pauli principle. This will lead to the construction of the *periodic system of elements* which is summarized in the *Mendeleev table*, first published in 1869 (much before quantum mechanics!) and based on empirical information[10]. The construction procedure is often referred to by the German word *aufbau* (which means 'building up').

Before proceeding, it is useful to recall and extend the nomenclature adopted in sections 3.1.4 and 3.5.5 for the spectroscopic classification of electronic states and shells. For most atoms[11] the single-electron energies E_{nl} follow this sequence of increasing value [2]

$$1s\ 2s\ 2p\ 3s\ 3p\ [3d \leftrightarrow 4s]\ 4p\ [5s \leftrightarrow 4d]\ 5p\ [6s \leftrightarrow 4f \leftrightarrow 5d]\ 6p\ [7s \leftrightarrow 5f \leftrightarrow 6d]$$

where levels in square parenthesis are so close in energy that, from case to case, they may be filled in the reverse order[12]. As already discussed for the hydrogen atom, the principal quantum number $n = 1, 2, 3,...$ defines the *electronic shell*, while the orbital quantum number $l = s, p, d, ...$ labels the *electronic sub-shells*. According to the Pauli principle, the maximum number of electrons that can be accommodated in a sub-shell with orbital quantum number l is $2(2l + 1)$. A sub-shell is said to be *incomplete* if it contains $k < 2(2l + 1)$ electrons and its configuration is indicated by the symbol $(nl)^k$. Finally, in total we can place up to $2n^2$ electrons in a shell with principal quantum number n.

A *closed shell* is a shell occupied by the maximum number of electrons it may accommodate: its corresponding quantum state is non-degenerate. On the other hand, the same electronic configuration of *an incomplete sub-shell can be realised in different ways*. Since within the central-field approximation the energy of each single-particle level is only assigned by the principal and angular quantum numbers, *there exists a degeneracy in the magnetic quantum numbers, both orbital and spin*. Therefore, if we have $k < 2n^2$ electrons, then there are

$$\frac{[2(2l + 1)]!}{k!\ [2(2l + 1) - k]!} \tag{5.54}$$

---

[10] Basically, Mendeleev investigated and organized the trends in the chemical behaviour of the elements.

[11] The sequence is obtained from spectroscopic evidence.

[12] For instance the sequence $[3d \leftrightarrow 4s]$ may appear inverted with respect to the hydrogen atom since $4s$ electrons can penetrate closer to the nucleus than the $3d$ ones and, therefore, Coulomb interactions become less screened or, equivalently, they are more strongly bound.

different ways of distributing them among all the possible single-electron states corresponding to the same set (nl) of quantum numbers. This results in some ambiguity in the definition of the ground state of atoms with an incomplete highest-energy sub-shell. This equivocation is resolved by the *empirical Hund rules* based on the definition of the *total orbital momentum* $\mathbf{L}_{tot}$ and *total spin momentum* $\mathbf{S}_{tot}$

$$\mathbf{L}_{tot} = \sum_{i=1}^{N} \mathbf{L}_i \quad \mathbf{S}_{tot} = \sum_{i=1}^{N} \mathbf{S}_i \tag{5.55}$$

where $\mathbf{L}_i$ and $\mathbf{S}_i$ are the angular and spin momentum of the ith electron, respectively. The quantum numbers associated with total momenta are l_{tot} and s_{tot}, respectively: they turn out to be 'good quantum numbers' for labelling states within the central-field approximation since it is easy to prove that

$$[\hat{H}_{cf}, \hat{\mathbf{L}}_{tot}] = 0 = [\hat{H}_{cf}, \hat{\mathbf{S}}_{tot}] \tag{5.56}$$

The Hund rules[13] *for incomplete sub-shells* containing $k < 2(2l + 1)$ equivalent[14] electrons are stated as follows:

1. the minimum-energy configuration has the maximum value of s_{tot};
2. for a given s_{tot}, the minimum-energy configuration has the maximum value of l_{tot};
3. for a given set of values for s_{tot} and l_{tot}, the minimum-energy configuration corresponds to $j_{tot} = l_{tot} - s_{tot}$ if $k < (2l + 1)$ or $j_{tot} = l_{tot} + s_{tot}$ if $k > (2l + 1)$.

Before proceeding, two comments should be made. First, in the third rule we have made use of the total angular momentum $\mathbf{J}_{tot} = \mathbf{L}_{tot} + \mathbf{S}_{tot}$, whose associated quantum number is j_{tot}. Next, we can say that the Hund rules represent a first fully empirical attempt to go beyond the independent electron picture. For instance, the first rule is the fingerprint of spin–spin interactions: a multi-electron configuration with a large s_{tot} corresponds to a larger number of electrons with parallel spin (unpaired electrons) than does a configuration with a small s_{tot} (corresponding to many paired electrons); these unpaired electrons repel each other (this notion has been learned in He and attributed to 'exchange forces'), thus lowering the total energy of the configuration. Similarly, the second and third rules are the manifestation of the orbit–orbit and spin–orbit interactions, although in a much less intuitive way. There is no need to insist on this argument, since all kinds of interactions will be explicitly treated in section 5.4.2. By applying in conjunction the Pauli principle and the Hund rules, we can determine the ground-state electronic configurations of all atoms, as reported in figure 5.6 for the first three shells with $n = 1$, 2, and 3. Although from this figure we can determine the configuration of only the first ten atoms, we can deduce some characteristics that apply to the entire periodic system of the elements. First of all, we remark that electrons belonging to

---

[13] *They represent the ultimate form of the rule anticipated in section 5.1.2.*
[14] In this framework, 'equivalent electrons' means: electrons with the same E_{nl} single-particle energy.

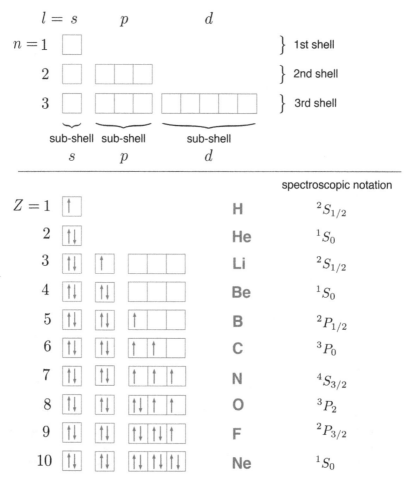

Figure 5.6. The *aufbau* of the ground-state electronic configuration of the first ten atoms of the periodic system. The top panel shows the graphical organization of electronic shells and sub-shells: each box represents a single-particle state characterized by the set of quantum numbers (nl); states belonging to the same multi-box have identical energy E_{nl} (in the central-field approximation). The bottom panel illustrates how the single-particle states are filled by electrons with either spin up ($\uparrow$) or spin down ($\downarrow$); the corresponding spectroscopic notation is shown on the right.

the highest-energy and incomplete sub-shell will be referred to as *valence electrons*, while electrons belonging to low-energy and complete sub-shells will be named *core electrons*. Next, we observe a regular repetition of the situation in which the electrons are in exact number to fill a certain number of sub-shells: this is the case of He, Be, and Ne, which correspond to the same spectroscopic symbol $^1S_0$ shown on the right[15]. Another regularity is found when atoms have the maximum number of electrons allowed in the shell: this is the case for He and Ne. Atoms with this

---

[15] The spectroscopic notation earlier introduced in equation (3.95) of section 3.5.5 for one-electron atoms is here extended to the case of multi-electron atoms by simply replacing L, S, J with their total counterparts.

property are named 'noble' since they are chemically inert: the full set contains He $(Z = 2)$, Ne $(Z = 10)$, Ar $(Z = 18)$, Kr $(Z = 36)$, Xe $(Z = 54)$, and Rn $(Z = 86)$. On the other hand, by adding one electron to the previous noble atom configuration we found a situation in which just the added electron falls outside a closed shell, as is found in Li. As we know already (see section 5.1.1), atoms of this kind are called 'alkali atoms' and their full series contains Li $(Z = 3)$, Na $(Z = 11)$, K $(Z = 19)$, Rb $(Z = 37)$, Cs $(Z = 55)$, and Fr $(Z = 87)$. Because of the outer lone s electron, alkali atoms are very chemically active. This property is similarly found in those atoms that are missing just one electron to completely fill the highest-energy sub-shell. In figure 5.6 this is the case for F $(Z = 9)$. They are customarily named 'halogens' and said to have a very large *affinity*, i.e., a very large tendency to attract an extra electron in the attempt to fill the incomplete sub-shell.

It is increasingly hard to proceed smoothly with the above procedure as soon as the principal quantum number becomes $n \geqslant 3$ since the sequence provided at the beginning of the present section departs from a plain ordering for increasing n. For instance, once the sub-shells $(3s)$ and $(3p)$ have been filled reaching the Ar $(Z = 18)$ atom, the next sub-shell actually filled is the $(4s)$ instead of the $(3d)$. The two atoms following Ar in the periodic system are K $(Z = 19)$ and Ca $(Z = 20)$ with, respectively, one and two $4s$ electron(s). After that, the $(3d)$ sub-shell is progressively filled, generating the 'first transition group' made by Sc, Ti, V, Cr, Mn, Fe, Co, Ni, Cu, and Zn with $21 \leqslant Z \leqslant 30$. The same situation occurs two more times, respectively after Kr $(Z = 36)$ and after Xe $(Z = 54)$, when the $[5s \leftrightarrow 4d]$ and the $[6s \leftrightarrow 5d]$ inversion occurs. These inversions generate the 'second transition group' and the 'third transition group', respectively corresponding to atoms with $39 \leqslant Z \leqslant 48$ and with $57 \leqslant Z \leqslant 80$.

Another inversion generates a different sequence of atoms named 'rare earths' (or lanthanides), found for elements with $58 \leqslant Z \leqslant 71$. In this case, the situation is determined by the $[4f \leftrightarrow 5d]$ swapping. Similarly, the 'actinides' group with $90 \leqslant Z \leqslant 103$ is determined by the competition $[5f \leftrightarrow 6d]$. Atoms with $Z > 100$ undergo spontaneous nuclear fission and, in particular for $Z > 103$, their electronic configuration is still largely unknown. Beyond $Z = 112$ the true existence of (meta) stable atoms is still a matter of discussion.

To a large extent *the chemical properties of an atom are dictated by its valence electrons*. It is therefore possible to organize the system of the elements so as to exploit those recurrences found in filling the highest-energy sub-shell. This leads to the construction of the *periodic table of the elements*, which is structured into columns of atoms with similar chemical properties (because they have incomplete and similarly-filled highest-energy sub-shells) and into rows of atoms, representing the progressive filling of the same sub-shell as Z increases. Columns are named *groups* and are 18 in total, while rows are named *periods* and are seven in total. This organization was first introduced by Mendeleev on the basis of phenomenological evidence and of an incomplete knowledge of existing elements, but it is still used by adding two extra-rows corresponding to the rare earths (or 'lanthanides') and to the actinides, respectively. Figure 5.7 reports the modern release of the Mendeleev periodic table. Nowadays the table is normally enriched by adding quite a lot of

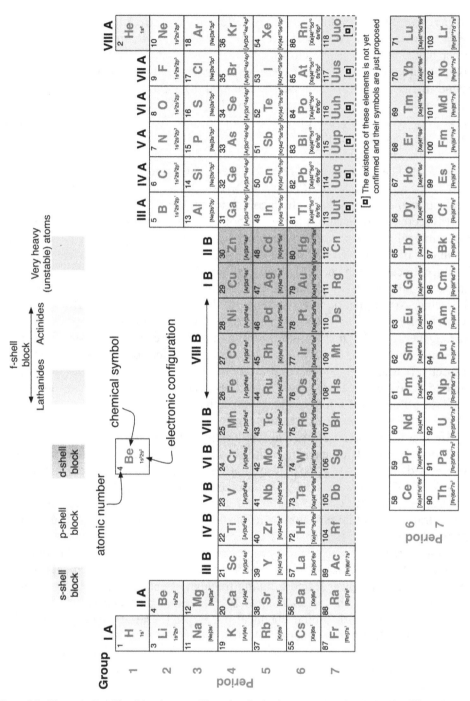

Figure 5.7. The periodic table of the elements. The color shadowing represents the progressive filling of *s*-, *p*-, *d*-, and *f*-shells. The last atomic elements of the seventh period are tentatively appointed as (increasing atomic number, starting from 113): Nihomium (Nh), Flevorium (Fl), Moscovium (Mc), Livermorium (Lv), Tennessine (Ts), and Oganesson (Og).

diverse information of chemical (atomic mass, atomic number, electronegativity, oxidation states, electronic structure, ionization energy, atomic radius, electron affinity, isotopic occurrence) or physical (mainly referred to the solid state like, e.g., crystal structure, density, melting temperature, molar volume, specific heat, electrical resistivity, thermal conductivity, melting enthalpy or even natural abundance on the Earth's crust) nature.

5.4 Beyond the central-field approximation

The central-field approximation, although corrected by the Pauli principle, turns out to be inadequate for quantitative predictions in a large number of interesting problems in atomic physics, beyond the *aufbau*. This is the case, for instance, for spectroscopy or astronomy where we are greatly interested in the optical transitions involving only those electrons accommodated in the highest-energy incomplete sub-shell[16]. In order to understand the experimental results for such transitions, we need to know in the most accurate way the actual sequence and the fine structure of the energy levels or, equivalently, it is necessary to go beyond an approximation where (i) electrons are treated as uncorrelated particles, (ii) it exists the exchange degeneracy, and (iii) spin-related features are only considered insofar as they determine, through the Pauli principle and the Hund rules, the accommodation of electrons on the different sub-shells. Interestingly enough, this fundamental knowledge is also greatly needed in molecular physics, in the chemical theory of interatomic bonding, and even in solid state physics.

The most rigorous way to accomplish this task, still preserving an independent electron picture, is to develop either the *Hartree (H) theory* or the *Hartree–Fock (HF) theory* which represent two somewhat similar theoretical models elaborated at different levels of sophistication. They form a very active field of research known as *quantum chemistry* [5, 6], together with many other approaches developed over the years and mostly based on numerical methods, including the more advanced *Configuration Interaction (CI) method*. The conceptual layouts of the H-, HF- and CI-methods are outlined below.

A less fundamental, but still very effective, way to proceed is based on the same *vector model of the atom* anticipated in section 3.4. Although this model is outclassed by the above quantitative methods, we will nevertheless treat it in detail since it is of great pedagogical value.

5.4.1 Hartree, Hartree–Fock, and Configuration Interaction methods

5.4.1.1 Problem definition, formal solution and notation
The common starting point of the H-, HF- and CI-methods is the Hamiltonian operator $\hat{H}$ given in equation (5.39), describing the electrostatic interactions within a system of spinless electrons. The goal we aim at is the *calculation of both the energy*

---

[16] Transitions between core states are also possible: they typically generate the absorption or emission of x-rays. An introductory discussion of the x-ray spectra is found elsewhere [2, 3].

E_{GS} *and the total wavefunction* Ψ_{GS} *of the ground state* of the atom described by this operator.

This task is accomplished by a variational approach [8], the rudiments of which were introduced in appendix G. More specifically, if Ψ is a trial total wavefunction, then the corresponding total energy $E[\Psi]$ must obey the condition

$$E_{GS} \leqslant E[\Psi] = \int \Psi^* \hat{H} \Psi \, dV \tag{5.57}$$

under the normalization constraint $\int \Psi^* \Psi dV = 1^{17}$. Formally this amounts to working out the minimization

$$E_{GS} = \min\{E[\Psi]\}_{\varphi} \tag{5.58}$$

with respect to variations of the *single-electron spin-orbitals* φ's.

The standard choice is to guess Ψ in the form of a Slater determinant

$$\Psi(1, 2,\ldots, N) = \frac{1}{\sqrt{N!}} \begin{vmatrix} \varphi_\alpha(1) & \varphi_\alpha(2) & \cdots & \varphi_\alpha(N) \\ \varphi_\beta(1) & \varphi_\beta(2) & \cdots & \varphi_\beta(N) \\ \cdots & \cdots & \cdots & \cdots \\ \varphi_\nu(1) & \varphi_\nu(2) & \cdots & \varphi_\nu(N) \end{vmatrix} \tag{5.59}$$

where the spin-orbitals are still unknown, but they are assumed to satisfy the normalization condition

$$\int_{m_{s_i}} \varphi_\beta^*(i)\varphi_\alpha(i) \, d\mathbf{r}_i = \delta_{\alpha\beta} \tag{5.60}$$

so that the Slater determinant does likewise. It is important to remark that the symbol $\int_{m_{s_i}} \cdots d\mathbf{r}_i$ indicates that two different operations are in fact executed for each single ith electron, namely (i) an integral over its space coordinates and (ii) a sum over its spin coordinate. This is a direct consequence of the fact that $\varphi_\alpha(i)$ are spin-orbitals.

5.4.1.2 The Hartree–Fock method

The key feature of the HF method is to consider separately the one-body (1b) and the two-body (2b) terms of equation (5.39): the first ones contain the contributions from the kinetic energy and from the electron–nucleus Coulomb interactions, the latter ones describe electron–electron coupling.

If we choose the trial HF total wavefunction Ψ_{HF} as in equation (5.59), this immediately leads to the definition of the *single-electron integrals* I_α^{1b} by setting

---

[17] For brevity we have set $dV = d\mathbf{r}_1 d\mathbf{r}_2 \cdots d\mathbf{r}_N$. Furthermore, in order to keep formalism as lean as possible, we will indicate the full set of quantum numbers describing a single-electron state by Greek letters $\alpha, \beta, \gamma,\ldots$ and the electron coordinates (both space and spin) by Latin letters $i, j \in [1, N]$.

$$\sum_{i=1}^{N} \int_{m_s} \Psi_{HF}^* \left[-\frac{\hbar^2}{2m_e}\nabla_i^2 - \frac{Ze^2}{4\pi\varepsilon_0}\frac{1}{r_i} \right] \Psi_{HF} \, dV = \sum_{\alpha} I_{\alpha}^{1b} \qquad (5.61)$$

where

$$I_{\alpha}^{1b} = \int_{m_{s_i}} \varphi_{\alpha}^*(i) \left[-\frac{\hbar^2}{2m_e}\nabla_i^2 - \frac{Ze^2}{4\pi\varepsilon_0}\frac{1}{r_i} \right] \varphi_{\alpha}(i) \, d\mathbf{r}_i \qquad (5.62)$$

The procedure to follow for the definition of the *two-electron integrals* is more complicated, but it proceeds similarly to the previous case. By setting

$$\sum_{i>j=1}^{N} \int_{m_s} \Psi_{HF}^* \left[\frac{e^2}{4\pi\varepsilon_0}\frac{1}{r_{ij}} \right] \Psi_{HF} \, dV = \frac{1}{2}\sum_{\alpha,\beta} [\underbrace{I_{\alpha\beta}^{2b,d}}_{\text{direct term}} - \underbrace{I_{\alpha\beta}^{2b,ex}}_{\text{exchange term}}] \qquad (5.63)$$

we calculate the *direct two-body integral*

$$I_{\alpha\beta}^{2b,d} = \frac{e^2}{4\pi\varepsilon_0} \int_{m_{s_i},m_{s_j}} \varphi_{\alpha}^*(i)\varphi_{\beta}^*(j) \frac{1}{r_{ij}} \varphi_{\alpha}(i)\varphi_{\beta}(j) \, d\mathbf{r}_i d\mathbf{r}_j \qquad (5.64)$$

and the *exchange two-body integral*

$$I_{\alpha\beta}^{2b,ex} = \frac{e^2}{4\pi\varepsilon_0} \int_{m_{s_i},m_{s_j}} \varphi_{\alpha}^*(i)\varphi_{\beta}^*(j) \frac{1}{r_{ij}} \varphi_{\beta}(i)\varphi_{\alpha}(j) \, d\mathbf{r}_i d\mathbf{r}_j \qquad (5.65)$$

While the $I_{\alpha\beta}^{2b,d}$ term is easily interpreted as the average Coulomb interaction in the two-electron state described by the wavefunction $\varphi_{\alpha}(i)\varphi_{\beta}(j)$, the exchange $I_{\alpha\beta}^{2b,ex}$ term has *no classical counterpart* since it directly derives from the determinant form of the total wavefunction.

By using the definitions given in equations (5.62), (5.64), and (5.65), we can express the total energy of the state Ψ_{HF} as

$$E[\Psi_{HF}] = \int_{m_s} \Psi_{HF}^* \hat{H} \Psi_{HF} \, dV = \sum_{\alpha} I_{\alpha}^{1b} + \frac{1}{2}\sum_{\alpha,\beta} \left[I_{\alpha\beta}^{2b,d} - I_{\alpha\beta}^{2b,ex} \right] \qquad (5.66)$$

The cumbersome minimization procedure defined in equation (5.58) under the normalization condition given in equation (5.60) eventually leads [8, 9] to the formidable set of integro-differential *HF equations*

$$\left[-\frac{\hbar^2}{2m_e}\nabla_i^2 - \frac{Ze^2}{4\pi\varepsilon_0}\frac{1}{r_i} \right]\varphi_{\alpha}(i) + \left[\sum_{\beta} \int_{m_{s_j}} \varphi_{\beta}^*(j)\frac{e^2}{4\pi\varepsilon_0}\frac{1}{r_{ij}}\varphi_{\beta}(j) \, d\mathbf{r}_j \right]\varphi_{\alpha}(i)$$
$$- \sum_{\beta}\left[\int_{m_{s_j}} \varphi_{\beta}^*(j)\frac{e^2}{4\pi\varepsilon_0}\frac{1}{r_{ij}}\varphi_{\alpha}(j) \, d\mathbf{r}_j \right]\varphi_{\beta}(i) = E_{\alpha}\varphi_{\alpha}(i) \qquad (5.67)$$

From the physical point of view, they are not ordinary eigenvalue problems since the potential energy terms depend on the very eigenfunctions representing the problem solutions. The HF equations are typically solved by numerical integration, through a self-consistent procedure (see section 5.2.1). The trial single-electron wavefunctions to start with are often guessed in the form of *Slater-type orbitals* whose normalized space part is written as [5, 7, 9]

$$\phi_{nlm}(\mathbf{r}) = \frac{(2\zeta)^{n+1/2}}{[(2n)!]^{1/2}} \, r^{n-1} \, \exp[-\zeta r] \, Y_{lm_l}(\theta, \phi) \tag{5.68}$$

where ζ is an adjustable parameter named *orbital exponent*, describing the radial decay of the wavefunction.

Since the HF equations are not Hamiltonian eigenvalue equations, *the quantity E_α cannot be interpreted as single-electron energy*. This issue is subtle: the sum $\sum_\alpha E_\alpha$ correctly counts once for each electron the contributions of its kinetic energy and its interaction with the nucleus, but it counts twice the mutual electron–electron interactions; therefore, $\sum_\alpha E_\alpha$ is not the total energy of the atom. We can nevertheless attach a very clean physical meaning to each energy E_α: the removal of the electron accommodated on the αth state of the initially neutral atom (in its ground state) costs an *ionization work W_α* which is straightforwardly calculated from equations (5.66) and (5.67)

$$W_\alpha = E^N[\Psi_{\mathrm{HF}}] - E^{N-1}[\Psi_{\mathrm{HF}}] = I_\alpha^{1b} + \sum_\beta \left[I_{\alpha\beta}^{2b,d} - I_{\alpha\beta}^{2b,ex} \right] = E_\alpha \tag{5.69}$$

where E^N and E^{N-1} are the total energy of the neutral and ionized atom, respectively. This conclusion[18] is called *Koopman's theorem* [5, 6], stating that *the quantity E_α represents the work needed to remove an electron from the state described by the spin-orbital φ_α*.

5.4.1.3 The Hartree method
The HF method is greatly simplified by *neglecting exchange terms*. This reduces equation (5.67) to a much simpler form

$$\left[-\frac{\hbar^2}{2m_e}\nabla_i^2 - \frac{Ze^2}{4\pi\varepsilon_0}\frac{1}{r_i} \right]\varphi_\alpha(i) + \left[\sum_{\beta \neq \alpha} \int_{m_{s_j}} \varphi_\beta^*(j)\frac{e^2}{4\pi\varepsilon_0}\frac{1}{r_{ij}}\varphi_\beta(j)\, d\mathbf{r}_j \right]\varphi_\alpha(i) = E_\alpha\varphi_\alpha(i) \tag{5.70}$$

They describe the classical situation of a system of interacting (and mutually overlapping) charge densities $-e|\varphi_\alpha(i)|^2$. Even in this case the solution is found

---

[18] We remark that, as highlighted by the notation used in equation (5.69), we are disregarding a subtle effect: upon the removal of one electron, the ground state wavefunction is affected by the readjustment of all remaining electrons. This is a real, but marginal effect, that we have neglected in our simplified formulation of the Koopman's theorem.

through an iterative self-consistent calculation, where the initial guess for the Hartree trial function Ψ_H is

$$\Psi_H = \varphi_\alpha(1)\varphi_\beta(2)\cdots\varphi_\nu(N) \tag{5.71}$$

This solution does not have the necessary antisymmetric property. Therefore, in order to enforce the Pauli principle, it must be understood that *only one electron can populate the state described by any given spin-orbital $\varphi_\alpha(i)$*. The missing antisymmetry is the consequence of neglecting the exchange effects: a choice, however, practiced whenever it is intended to minimize the computational cost of the self-consistent procedure, which in this case is much lighter than in the HF method just because of the missing exchange term.

5.4.1.4 The Configuration Interaction method
Instead of simplifying the HF theory as in the Hartree method, we could aim at developing an even more rigorous solution. In fact, the energy given in equation (5.66) is just an excess approximation to the true total energy E_{GS}^{true}, being the result of a variational procedure for which it holds equation (5.58). The difference $E_c = E_{GS}^{true} - E[\Psi_{HF}]$ is called *correlation energy*: this is the quantity to be addressed in our attempt to go beyond HF theory.

There basically exist two approaches to calculate the correlation energy: either (i) to treat it as perturbation on the HF solutions or (ii) to cast the trial wavefunction in the form of a *linear combination of Slater determinants*, where different sets of spin-orbitals are used in each determinant. This second option is called the CI method [5], whose development falls beyond the scopes of the present Primer.

We finally remark that the accurate evaluation of both the exchange and the correlation effects in multi-electron systems (atoms, molecules, and solids) is a very active and still open problem in many-body physics.

5.4.2 The vector model: L–S and J–J coupling schemes

5.4.2.1 The model formulation
As already discussed, the basic idea of the vector model of the atom consists in describing any angular momentum (both orbital and spin) at first as an ordinary vector in classical mechanics, in particular as regards the composition (sum or subtraction) rules. Next, quantum features are invoked by using the quantum expectation values for the resulting vectors. Finally, we recognise that any moment is associated with a magnetic dipole, which we will describe by extending the arguments developed in sections 3.3 and 3.4.

In practice, starting from the central field picture, we introduce the following couplings for any electron pair i, j

$$\text{orbit} - \text{orbit coupling}: \quad \xi_{LL} \, \mathbf{L}_i \cdot \mathbf{L}_j$$
$$\text{spin} - \text{spin coupling}: \quad \xi_{SS} \, \mathbf{S}_i \cdot \mathbf{S}_j \tag{5.72}$$
$$\text{spin} - \text{orbit coupling}: \quad \xi_{LS} \, \mathbf{L}_i \cdot \mathbf{S}_j$$

The spin–spin interaction is nothing other than the exchange interaction described in section 5.1.2 and, therefore, we can assume $\xi_{SS} = -2K$, where $K > 0$ is a suitable exchange integral that, in a first instance, can be calculated by using central-field wavefunctions. On the other hand, the spin–orbit term is treated similarly to the theory developed in section 3.4; the corresponding ξ_{LS} coupling constant is calculated as in equation (3.81) by replacing the Coulomb potential $V(r)$ with the central-field potential $V_{cf}(r)$. The ξ_{LL} coupling constant is much less straightforward and we will consider it as a phenomenological constant to be empirically determined. The three coupling constants have different sign: while ξ_{LL} and ξ_{SS} are always negative, the spin–orbit term ξ_{LS} could be either positive and negative, depending on the number of electrons, as established by this twofold rule of thumb: (i) in general, in the lowest-energy state the k electrons not belonging to complete sub-shells are accommodated on the same (nl) level; and (ii) it results in $\xi_{LS} > 0$ or $\xi_{LS} < 0$ provided that $2k$ is, respectively, smaller or larger than the maximum number of electrons that could be accommodated on this level[19]. Finally, in order to calculate the scalar products between angular momenta we can adopt the same procedure used in sections 3.4 and 5.1.2. The resulting picture is sketched in figure 5.8 for a simple two-electron system[20], highlighting the fact that orbital and intrinsic angular momenta generate corresponding magnetic dipoles $\mathbf{M}_{L_i}$ and $\mathbf{M}_{S_i}$ which, in turn, mutually interact through the couplings provided by equation (5.72). Overall, the major improvement with respect to the central-field approximation consists in the explicit calculations of the *magnetic interactions*, previously neglected.

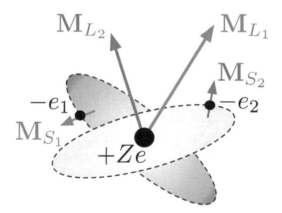

Figure 5.8. Schematic representation of an atom with only two electrons in the highest-energy incomplete sub-shell. Their orbital and magnetic momenta, respectively $\mathbf{M}_{Li}$ and $\mathbf{M}_{Si}$ (with $i = 1, 2$), are explicitly shown.

---

[19] This rule of thumb can be derived from the Hund rules [4].
[20] We remark once again that we are addressing only electrons belonging to the highest-energy incomplete sub-shell, since it holds that $\mathbf{L}_{tot} = \mathbf{S}_{tot} = 0$ for electrons in a complete sub-shell.

It is experimentally observed that such couplings have different relative strength as Z increases: in lighter atoms the spin–orbit coupling is the smallest one, while in large-Z atoms it dominates over orbit–orbit and spin–spin ones. These two different situations can be separately treated in the so-called *L–S (or Russell–Saunders) scheme and J–J scheme* [4, 10], respectively, as discussed below.

5.4.2.2 The L–S coupling scheme
For small enough Z values, it generally holds that $|\xi_{SS}|, |\xi_{LL}| > |\xi_{LS}|$. If just two electrons are accommodated in the highest-energy incomplete sub-shell, then we can write

$$\mathbf{L}_{tot} = \mathbf{L}_1 + \mathbf{L}_2 \quad L_{tot} = \sqrt{l_{tot}(l_{tot} + 1)}\ \hbar \tag{5.73}$$

$$\mathbf{S}_{tot} = \mathbf{S}_1 + \mathbf{S}_2 \quad S_{tot} = \sqrt{s_{tot}(s_{tot} + 1)}\ \hbar \tag{5.74}$$

so that the *total angular momentum* $\mathbf{J}_{tot}$ is defined as

$$\mathbf{J}_{tot} = \mathbf{L}_{tot} + \mathbf{S}_{tot} \quad J_{tot} = \sqrt{j_{tot}(j_{tot} + 1)}\ \hbar \tag{5.75}$$

We can model the situation by saying that the stronger orbit–orbit and spin–spin interactions act so as to generate, respectively, the total angular and spin momentum: because of the spin–orbit coupling, they both precess around the direction of $\mathbf{J}_{tot}$. In order to determine all possible values of the quantum numbers l_{tot}, s_{tot}, and j_{tot} we use the classical rules for the vector composition. As an example, let us suppose that the two electrons of figure 5.8 are a (d) and a (p) one, respectively. As regards the total spin quantum number s_{tot} we have just two possibilities, namely

$$\text{spin composition:} \begin{cases} \uparrow + \uparrow & \rightarrow s_{tot} = 1 \\ \uparrow + \downarrow & \rightarrow s_{tot} = 0 \end{cases} \tag{5.76}$$

while for the total angular quantum number l_{tot} we can calculate

$$\text{angular momentum composition:} \begin{cases} d + p & \rightarrow \text{maximum value } l_{tot} = 3 \\ d - p & \rightarrow \text{minimum value } l_{tot} = 1 \end{cases} \tag{5.77}$$

Since the possible values of the quantum number associated with any angular momentum must differ by one $\hbar$ unity, from equation (5.77) we obtain that the orbital quantum number can assume the values $l_{tot} = 1, 2, 3$. The above two equations define all possible spin–spin and orbit–orbit couplings.

We now turn to evaluate the spin–orbit interaction, which is of course null for the $s_{tot} = 0$ case. On the other hand, when $s_{tot} = 1$ we must distinguish three different cases

$$s_{tot} = 1 \text{ with } l_{tot} = 3 \rightarrow \left\{ \begin{array}{l} \text{maximum } j_{tot} \text{ value: } 3+1=4 \\ \text{minimum } j_{tot} \text{ value: } 3-1=2 \end{array} \right\} \rightarrow j_{tot} = 4, 3, 2$$

$$s_{tot} = 1 \text{ with } l_{tot} = 2 \rightarrow \left\{ \begin{array}{l} \text{maximum } j_{tot} \text{ value: } 2+1=3 \\ \text{minimum } j_{tot} \text{ value: } 2-1=1 \end{array} \right\} \rightarrow j_{tot} = 3, 2, 1 \quad (5.78)$$

$$s_{tot} = 1 \text{ with } l_{tot} = 1 \rightarrow \left\{ \begin{array}{l} \text{maximum } j_{tot} \text{ value: } 1+1=2 \\ \text{minimum } j_{tot} \text{ value: } 1-1=0 \end{array} \right\} \rightarrow j_{tot} = 2, 1, 0$$

The states corresponding to the same set of quantum numbers (l_{tot}, s_{tot}) and differing by j_{tot} are said to form a *multiplet*: they are resolved (i.e. split) by the spin–orbit interaction. The fine structure for our benchmark case shown in figure 5.9 is eventually obtained by combining equations (5.76), (5.77), and (5.78). We remark that whenever the incomplete sub-shell contains two equivalent electrons (i.e., two electrons with the same orbital quantum number) some additional care must be played when developing the vector composition [2].

The spin–orbit coupling deserves some additional comment. Its corresponding interaction energy E_{so} is cast in the same form as in equation (3.88)

$$E_{so} = \frac{\xi_{LS}}{2} \left[j_{tot}(j_{tot} + 1) - l_{tot}(l_{tot} + 1) - s_{tot}(s_{tot} + 1) \right] \quad (5.79)$$

While, as anticipated, the calculation of ξ_{LS} is more complicated than in the hydrogenic case because of the actual form of the central-field self-consistent potential, its value is just the same for any level with the identical l_{tot} and s_{tot} quantum numbers, as experimentally confirmed: the spin–orbit separation ΔE_{so} between any pair of adiacent fine structure levels, i.e., for any pair of levels with total angular quantum numbers j_{tot} and ($j_{tot} + 1$), is given by

$$\Delta E_{so} = \xi_{LS} (j_{tot} + 1) \quad (5.80)$$

which is known as the *Landé interval rule*[21]. More generally, if $\xi_{LS} > 0$ the lower state of a given multiplet has the minimum value of j_{tot}: the multiplet is said to be *regular*. In the opposite case $\xi_{LS} < 0$ the multiplet is named *inverted*, meaning for the maximum j_{tot} value we find the minimum possible energy.

Finally, we observe that if the highest-energy sub-shell contains more than two electrons, everything is conceptually similar to the above worked benchmark case, although the practical use of the rules for the vector composition are somewhat more complicated: initially, we need to compose the angular momenta of the first two electrons; next, we must further compose the resulting momentum with that one of the third electron, and so on. The procedure is illustrated in the case of three electrons in figure 5.10.

---

[21] More rigorously, the Landé interval rules holds for those light atoms for which the L–S coupling scheme can be applied, but not for very light atoms: here the relativistic effects are so strong that deviations from the interval rule are in fact observed.

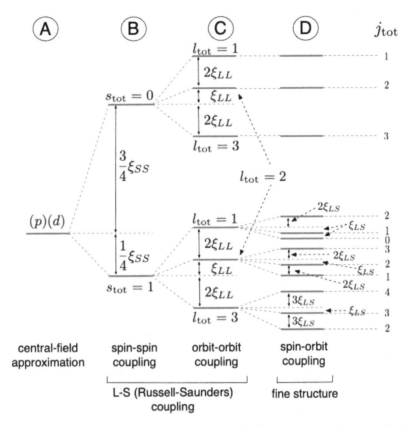

Figure 5.9. The fine structure of a hypothetical incomplete sub-shell containing one (p) and one (d) electron in an atom with small Z. (A) The degenerate energy level obtained within the central-field approximation. (B–D) The spin–spin, orbit–orbit, and spin–orbit coupling is added, respectively. The structure provided by panels (B) and (C) is known as the L–S (Russell–Saunders) coupling scheme, while panel (D) describes the fine structure of the spectrum (i. e. it shows the relativistic effects). The ξ's coupling constants are defined in equation (5.72). In plotting this picture it has been assumed $\xi_{SS}, \xi_{LL} < 0$ with $|\xi_{SS}| > |\xi_{LL}|$ and $\xi_{LS} > 0$ (regular spin–orbit multiplet). On the right column states are classified according to their total angular momentum quantum number j_{tot}.

5.4.2.3 The J–J coupling scheme

In heavy atoms the strong spin–orbit interaction tightly couples the orbital momentum and the spin momentum of each ith electron, thus determining $\mathbf{J}_i = \mathbf{L}_i + \mathbf{S}_i$. The remaining electron–electron interactions determine the resulting $\mathbf{J}_{\text{tot}} = \sum_i \mathbf{J}_i$. By considering once again the case of an incomplete sub-shell containing just two electrons, the quantum number j_{tot} is defined in the interval $|j_1 - j_2| \leqslant j_{\text{tot}} \leqslant j_1 + j_2$, the possible values being separated by one $\hbar$ unity.

If we indicate by $\xi_{LS,i}$ the spin–orbit coupling constant for the ith electron and by ξ_{JJ} the electron–electron coupling constant, then the energies $E_{\text{tot,cf}}$ obtained within the central-field approximation must be corrected by adding (i) a $\sum_i \xi_{LS,i} \mathbf{L}_i \cdot \mathbf{S}_i$ contribution taking care of the complete spin–orbit interaction, and (ii) a

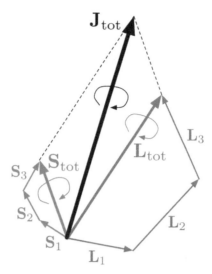

Figure 5.10. The L–S coupling scheme applied to a three-electron system to obtain $\mathbf{L}_{tot}$, $\mathbf{S}_{tot}$, and $\mathbf{J}_{tot}$. The arrowed closed loops pictorially indicate that the component vectors precess around the direction of their sum.

$\xi_{JJ} \sum_{i \neq k} \mathbf{J}_i \cdot \mathbf{J}_k$ contribution overall describing any remaining electron–electron interaction.

If we reconsider once again our benchmark case of an incomplete sub-shell containing just two non-equivalent electrons, then its fine structure is given by

$$E_{JJ-\text{coupling}} = E_{\text{tot,cf}}$$

$$+ \underbrace{\sum_{i=1}^{2} \frac{\xi_{LS,i}}{2} [j_i(j_i + 1) - l_i(l_i + 1) - s_i(s_i + 1)]}_{\text{spin–orbit interactions}} \qquad (5.81)$$

$$+ \underbrace{\frac{\xi_{JJ}}{2} [j_{\text{tot}}(j_{\text{tot}} + 1) - j_1(j_1 + 1) - j_2(j_2 + 1)]}_{\text{remaining electron–electron interactions}}$$

In figure 5.11 we show the fine structure in the case of two non-equivalent (s) and (p) electrons. This is a particularly simple case since one electron (conventionally labelled as the '1' electron) lies in an s-state and, therefore, it does not feel the spin–orbit interaction.

5.5 Multi-electron atoms in excited states

A multi-electron atom can be excited from its ground state by collisions with other atoms or electrons, as well as by absorbing photons with suitable energy. We can basically distinguish between *single-electron excitations* and *multi-electron excitations*, by considering the number of particles undergoing a transition. A common feature is that any such excitation indeed affects all the remaining electrons, because of the correlation existing among them: exciting one (or more) member(s) of the

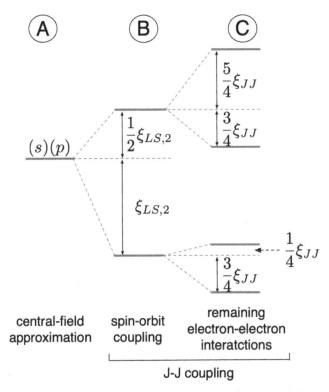

Figure 5.11. The fine structure of a hypothetical incomplete sub-shell containing one (s) and one (p) electron in an atom with large Z. (A) The degenerate energy level obtained within the central-field approximation. (B–C) The spin–orbit and the remaining (i. e. not included in the central-field approximation) electron–electron couplings are added, respectively. The structure provided by panels (B) and (C) is known as J–J coupling scheme. In plotting this picture it has been assumed $\xi_{LS,2} > \xi_{JJ} > 0$ (where '2' is the electron in the p-state).

electron family effectively changes the overall set of Coulomb interactions[22]. It is easy to justify this statement within the central field approximation: the actual form of V_{cf} is changed by varying even just one single-electron wavefunction (namely, that corresponding to the excited electron), as made clear by inspection of equation (5.49). This means that *the energy absorbed by the atom through a collision or by absorbing a photon is in part transferred to the excited electron(s) and in part distributed among the other ones.*

In single-electron excitations, a comparatively smaller energy is required to promote a valence electron to a higher state, with respect to the case in which a core electron is excited from a inner shell. A typical excitation energy of 1–10 eV or 10–10^4 eV is requested in the two cases, respectively[23]. In addition, when a core

---

[22] The same is also true for the magnetic interactions; however, since the effect is much smaller, we will neglect them in our qualitative discussion.

[23] This rule of thumb is not valid for noble gases, for which the ground state corresponds to a completely filled highest-energy shell: for them the single-electron excitation energy can be as large as some tens of eV even if the promoted electron was initially accommodated on the outer shell.

electron is excited, a 'hole' state (that is, a 'missing electron' situation) is created which is available to receive a second electron, decaying from some higher level. The energy released during this decaying process typically corresponds to an x-ray photon: in fact, this physical mechanism is used in x-ray tubes [3]. Alternatively, the same energy release can promote the extraction of a different electron to an unbound state, thus generating an auto-ionization process of the atom which is usually referred to as *Auger effect*. Both processes are pictorially illustrated in figure 5.12.

Multi-electron processes are much more complicated and, therefore, we limit ourselves to mentioning just the case of two electrons simultaneously excited. The corresponding energy balance is subtle since it is given by the sum of three contributions, namely: the two energies required to promote each electron individually, plus the change in the interaction energy of the whole electron family caused by the double excitation. The associated phenomenology is correspondingly complex, since it can take different forms, like: (i) the two excited electrons can decay by *emitting two photons* (generally of different energy); (ii) the non-radiative release of energy from one of the two electrons can further excite the other, consequently promoting it to an even higher state in energy. In this case, the *auto-ionization of the atom* is observed provided that the final state of the doubly-excited electron falls beyond the ionization threshold. The rate of occurrence for two-electron processes (both photons emission and non-radiative auto-ionization) anyway critically depends on the principal quantum numbers of the excited states these very electrons are promoted to: the larger these numbers, the less likely is the process. We can easily understand this by considering the fact that by increasing the principal quantum numbers the average electron–electron distance increases accordingly: therefore, their mutual interaction weakens and the lifetime of the two excited states increases accordingly (which makes the following evolution less probable).

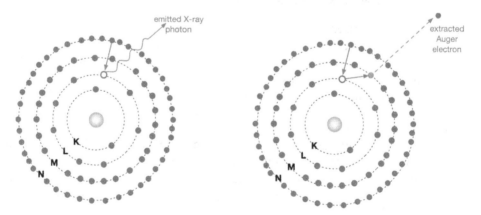

Figure 5.12. Schematic representation of one-electron excitation processes in a multi-electron atom. Left: emission of an x-ray photon by the refilling of a 'hole' state previously generated. Right: electron extraction upon energy release by the refilling of a 'hole' state (Auger effect). Electronic shells ($2n^2$-fold degenerate) are labelled K, L, M, N, ... as described in section 3.1.4 and pictorially represented by dashed lines; electrons are shown by full blue dots and the 'hole' generated by a previous excitation is shown as an empty blue circle. Red and magenta lines highlight, respectively, the primary electron decay and the associated physical process.

5.6 Selection rules

Optical transitions are possible between two levels of the energy spectrum of a multi-electron atom, obeying selection rules similarly to the case of hydrogenic atoms discussed in section 4.3. They are obtained by evaluating the electric dipole matrix element, a task falling beyond the knowledge developed so far (details can be found in [2, 3, 10]); here we limit ourselves to reporting the main results.

In both the L–S and J–J coupling scheme, an electric dipole transition must fulfil the condition

$$\Delta j_{tot} = 0, \pm 1 \tag{5.82}$$

as regards the total angular momentum quantum number; however, it must be understood that the transition $j_{tot} = 0 \rightarrow j_{tot} = 0$ is in any case forbidden. When an external magnetic field is applied so as to split levels with different $m_{j_{tot}}$ values, it must also obey the condition

$$\Delta m_{j_{tot}} = 0, \pm 1 \tag{5.83}$$

In any case, the single electron undergoing the transition is governed by the additional rule $\Delta l = \pm 1$ in the L–S scheme and by the rules $\Delta j = 0, \pm 1$ and $\Delta l = \pm 1$ in the J–J scheme. Higher-order (magnetic dipole of electric quadrupole) transitions are also possible [2].

5.7 The action of an external magnetic field

We conclude our journey in the physics of multi-electron atoms by considering the effect of an external, static and uniform magnetic field **B**. Since the internal fields sum up to ~1 T, we will treat separately the two limiting situations of *weak external field* $|\mathbf{B}| < 1$ T (Zeeman effect) and of *strong external field* $|\mathbf{B}| > 1$ T (Paschen–Back effect).

In the *Zeeman case* (shown in figure 5.13 for the Na atom), the total magnetic moment of the atom is

$$\mathbf{M}_{J_{tot}} = \mathbf{M}_{L_{tot}} + \mathbf{M}_{S_{tot}} = -g_L \frac{\mu_B}{\hbar} \mathbf{L}_{tot} - g_S \frac{\mu_B}{\hbar} \mathbf{S}_{tot} \tag{5.84}$$

where the total angular momenta and their relative quantum numbers are defined in equation (5.55), while $g_L = 1$ and $g_S = 2$. Under the action of the external field we observe (i) the precession of $\mathbf{L}_{tot}$ and $\mathbf{S}_{tot}$ around $\mathbf{J}_{tot}$, and (ii) the precession of $\mathbf{J}_{tot}$ around **B**. Therefore, the net contribution of both the orbital and the spin angular momentum are only provided by their components along $\mathbf{J}_{tot}$ which we, respectively, indicate by $\mathbf{L}_{tot,J_{tot}}$ and $\mathbf{S}_{tot,J_{tot}}$. More specifically, we can write

$$\mathbf{L}_{tot,J_{tot}} = \left(\mathbf{L}_{tot} \cdot \frac{\mathbf{J}_{tot}}{J_{tot}} \right) \frac{\mathbf{J}_{tot}}{J_{tot}} = \left(\frac{\mathbf{L}_{tot} \cdot \mathbf{J}_{tot}}{J_{tot}^2} \right) \mathbf{J}_{tot} \tag{5.85}$$

and

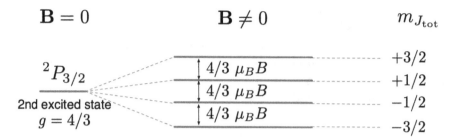

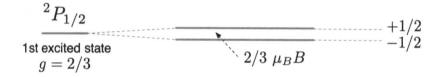

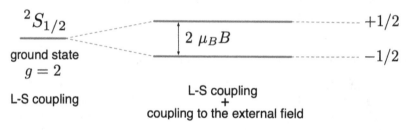

Figure 5.13. The Zeeman effect in sodium. Note that the splitting is different, according to the different value assumed by the Landé g-factor g.

$$\mathbf{S}_{\text{tot},J_{\text{tot}}} = \left(\mathbf{S}_{\text{tot}} \cdot \frac{\mathbf{J}_{\text{tot}}}{J_{\text{tot}}} \right) \frac{\mathbf{J}_{\text{tot}}}{J_{\text{tot}}} = \left(\frac{\mathbf{S}_{\text{tot}} \cdot \mathbf{J}_{\text{tot}}}{J_{\text{tot}}^2} \right) \mathbf{J}_{\text{tot}} \tag{5.86}$$

The *net atomic magnetic moment along* $\mathbf{J}_{\text{tot}}$ is straightforwardly calculated to be

$$\mathbf{M}_{J_{\text{tot}}} = -\frac{\mu_{\text{B}}}{\hbar J_{\text{tot}}^2} (\mathbf{L}_{\text{tot}} \cdot \mathbf{J}_{\text{tot}} + 2\,\mathbf{S}_{\text{tot}} \cdot \mathbf{J}_{\text{tot}}) \mathbf{J}_{\text{tot}}$$

$$= -\frac{\mu_{\text{B}}}{\hbar} g\, \mathbf{J}_{\text{tot}} \tag{5.87}$$

where the coefficient

$$g = 1 + \frac{j_{\text{tot}}(j_{\text{tot}} + 1) - l_{\text{tot}}(l_{\text{tot}} + 1) + s_{\text{tot}}(s_{\text{tot}} + 1)}{2j_{\text{tot}}(j_{\text{tot}} + 1)} \tag{5.88}$$

is named *Landé g-factor* for the multi-electron atom.

When the magnetic dipole described by equation (5.87) is subjected to an external field $\mathbf{B}$ a new term

$$E_{\text{mag}} = \mu_B \, g \, m_{J_{\text{tot}}} \, B \tag{5.89}$$

must be added to the multi-electron energy spectrum: this causes the split of the levels degenerate in $m_{J_{\text{tot}}}$. Experimentally the Zeeman effect is observed by looking for the optical transitions occurring between the split levels and obeying the selection rules discussed in the previous section: the action of the external magnetic field is easily recognized since each single spectral line observed in the absence of any field is now resolved into a more complex pattern, containing several components. In order to illustrate this feature we show in figure 5.13 the case of the first three states of the Na atom.

In the *Paschen–Back* case, the total orbital and the spin momentum undergo independent precessions around the direction of the external field $\mathbf{B}$, resulting in a magnetic coupling energy

$$E_{\text{mag}} = \frac{\mu_B B}{\hbar}(L_{\text{tot},B} + 2S_{\text{tot},B}) = \mu_B B \, (m_{l_{\text{tot}}} + 2m_{s_{\text{tot}}}) \tag{5.90}$$

where $L_{\text{tot},B}$ and $S_{\text{tot},B}$ are the vector components along the direction of the magnetic field. It must be noted that in this case the selection rule $\Delta m_{s_{\text{tot}}} = 0$ must be obeyed [2].

Further reading and references

[1] Colombo L 2021 *Solid State Physics: A Primer* (Bristol: IOP Publishing)
[2] Bransden B H and Joachain C J 1983 *Physics of Atoms and Molecules* (Harlow: Addison-Wesley Longman)
[3] Demtröder W 2010 *Atoms, Molecules and Photons* (Heidelberg: Springer)
[4] Rigamonti A and Carretta P 2009 *Structure of Matter* 2nd edn (Milano: Springer)
[5] Szabo A and Ostlund N S 1996 *Modern Quantum Chemistry* (New York: Dover)
[6] Atkins P and Friedman R 2011 *Molecular Quantum Mechanics* 5th edn (Oxford: Oxford University Press)
[7] Slater J C 1968 *Quantum Theory of Matter* (New York: McGraw-Hill)
[8] Messiah A 1968 *Quantum Mechanics* (Amsterdam: North-Holland)
[9] Atkins P and De Paula J 2010 *Physical Chemistry* 9th edn (Oxford: Oxford University Press)
[10] Foot C J 2005 *Atomic Physics* (Oxford: Oxford University Press)

Part III

Molecular physics

IOP Publishing

Atomic and Molecular Physics (Second Edition)
A primer
Luciano Colombo

Chapter 6

Molecules: general features

Syllabus—In this chapter we discuss two important introductory topics to molecular physics, namely: the Born–Oppenheimer approximation and the nature of the interatomic bond. In particular, after defining the general problem we will discuss how, and under what conditions, it is possible to separate nuclear and electronic motions. This separation is the basis of the independent treatments of the roto-vibrational properties (chapter 7) and of the electronic structure (chapter 8) of molecules. Lastly, the physics underlying the formation of interatomic bonds will be discussed in the two paradigmatic cases of diatomic molecules characterized by ionic and covalent bonding, respectively.

6.1 What is a molecule?

Molecules are the smallest part of a compound substance able to fully define its chemical properties. They are made by two or more atoms (even very many, as in the case of biological molecules) and, therefore, at the most fundamental level they can be defined as *a bound system of nuclei and electrons.*

This very simple definition hides a number of subtle key issues.

First of all, let us dwell on this concept: the definition given above suggests that a molecule should be looked at as *an assembly of nuclei, surrounded by electrons.* This picture is surely the most appropriate one if we are interested in quantitatively predicting the molecular electronic configurations, as needed e.g., to correctly interpret the molecular optical spectra. Alternatively, we could assume that *atoms forming a molecule maintain their own individuality.* Even this rather different picture is in fact correct: the empirical evidence indicates that most of the chemical properties of a molecule are primarily determined by the valence electrons of its constituent atoms which, therefore, must be considered as individual entities. Similarly, the physics of molecular vibrations and rotations is effectively described by a model where constituent atoms mutually exchange mechanical actions through

doi:10.1088/978-0-7503-5734-0ch6

specific interatomic forces. In summary, quite rightly atoms can be considered as the building blocks of a molecule.

Next, we remark that the complete problem in molecular physics is defined in terms of both nuclear and electronic degrees of freedom. In other words: in principle, *a molecule is not two separate roto-vibrational and electronic-structure problems.* Therefore, we are challenged by a fundamental question: at what level and how do electronic and nuclear motions affect each other? Could we separate them, or are we rather forced to develop a theory where nuclear and electronic features are inherently entangled?

This brief introduction defines the twofold goal of the present chapter: while the next section is addressed to elaborate the most appropriate picture to investigate the physics of a molecule, the following one outlines a phenomenological theory of molecular cohesion.

6.2 The Born–Oppenheimer approximation

The formulation of the quantum mechanical molecular problem is tricky and, therefore, our first step must consist in defining a suitable notation. Nuclear positions will be labeled by Greek letters and indicated by the capital symbol $\mathbf{R}_\alpha$; on the other hand, we will, respectively, use Latin letters and the lowercase symbol $\mathbf{r}_i$ for electrons. Whenever we must indicate the *full set of nuclear or electronic coordinates*, we will use the shortcut $\mathbf{R}$ or $\mathbf{r}$, respectively.

By following a similar approach extensively used for atoms, *we neglect in the first instance any magnetic and relativistic effect.* The consequence of this assumption is twofold. On the one hand, the *total molecular wavefunction* only depends on the space coordinates of electrons and nuclei (i.e., no spin degrees of freedom at this level): it will be indicated as $\Psi(\mathbf{r}, \mathbf{R})$. On the other hand, only Coulomb interactions are considered, which are of three kinds (classical expressions are provided in the next three equations):

- the *nucleus–electron coupling*

$$V_{\text{ne}}(\mathbf{r}, \mathbf{R}) = -\frac{1}{4\pi\varepsilon_0} \sum_{i,\alpha} \frac{Z_\alpha e^2}{|\mathbf{r}_i - \mathbf{R}_\alpha|} \tag{6.1}$$

where the sum runs over all electron and nuclear positions, while $+Z_\alpha e$ is the charge of the αth nucleus;

- the *nucleus–nucleus coupling*

$$V_{\text{nn}}(\mathbf{R}) = \frac{1}{4\pi\varepsilon_0} \sum_{\alpha > \beta} \frac{Z_\alpha Z_\beta e^2}{|\mathbf{R}_\alpha - \mathbf{R}_\beta|} \tag{6.2}$$

where the sum runs over all nuclear pairs;

- the *electron–electron coupling*

$$V_{\text{ee}}(\mathbf{r}) = \frac{1}{4\pi\varepsilon_0} \sum_{i > j} \frac{e^2}{|\mathbf{r}_i - \mathbf{r}_j|} \tag{6.3}$$

where the sum runs over all electron pairs.

In this formal framework the *non-relativistic energy eigenvalue equation for a generic molecule* is written as

$$\left[-\frac{\hbar^2}{2}\sum_\alpha \frac{1}{M_\alpha}\nabla_\alpha^2 - \frac{\hbar^2}{2m_e}\sum_i \nabla_i^2 + \hat{V}_{ne}(\mathbf{r},\mathbf{R}) + \hat{V}_{nn}(\mathbf{R}) + \hat{V}_{ee}(\mathbf{r}) \right]\Psi(\mathbf{r},\mathbf{R}) = E_T\Psi(\mathbf{r},\mathbf{R}) \quad (6.4)$$

where M_α is the mass of the αth nucleus and E_T the *total molecular energy*, including any kind of nuclear and electronic contribution. In this equation the symbols ∇_α^2 and ∇_i^2 imply, respectively, derivatives with respect to nuclear and electronic coordinates, while $\hat{V}_{ne}$, $\hat{V}_{nn}$, and $\hat{V}_{ee}$ are the operator counterparts of the three Coulomb potential energies.

This problem has no analytical solution: simply, too many degrees of freedom and too many entangled interactions. Nevertheless, the experimental evidence points our path in the right direction: molecular optical spectra are characterized by absorption/emission lines which are easily grouped in series falling in quite diverse regions of the electromagnetic spectrum [1–4]. While transitions involving electronic states fall between the visible (VIS) and the ultraviolet (UV) regions, transitions involving nuclear motions are found in between the infrared (IR) and the microwave (MW) regions. More specifically, molecular vibrations and rotations, respectively, involve emission/absorption of IR and MW photons. This clearly indicates that *the physics of electrons and nuclei in a molecule unfolds on very different energy scales* and, therefore, we can imagine decoupling and treating them separately: this is indeed a great simplification, usually referred to as *Born–Oppenheimer approximation* (or, as explained below, *adiabatic approximation*).

Just as space and spin coordinates in atomic physics can be treated separately because the corresponding interactions have very different energetics, *we argue that molecular degrees of freedom can be similarly separated*. We therefore exploit the Born–Oppenheimer approximation by writing the total molecular wavefunction in the form

$$\Psi(\mathbf{r},\mathbf{R}) = \psi_n(\mathbf{R})\,\psi_e^{(\mathbf{R})}(\mathbf{r}) \quad (6.5)$$

where $\psi_n(\mathbf{R})$ and $\psi_e^{(\mathbf{R})}(\mathbf{r})$ are, respectively, the *nuclear wavefunction* and the *electronic wavefunction*. The first one is in charge of describing any kind of roto-vibrational motion of the molecule: it only depends on the nuclear coordinates and does not carry any information about the electronic configuration which, in turn, is fully described by the second function. More precisely: $\psi_e^{(\mathbf{R})}(\mathbf{r})$ *describes the electron states for each clamped-nuclei configuration* $\mathbf{R}$. This implies that it has an *analytical dependence* on the $\mathbf{r}$-coordinates and a *parametric dependence* on the $\mathbf{R}$-coordinates. It is obtained by solving the eigenvalue problem

$$\left[-\frac{\hbar^2}{2m_e}\sum_i \nabla_i^2 + \hat{V}_{ne}(\mathbf{r},\mathbf{R}) + \hat{V}_{ee}(\mathbf{r}) \right]\psi_e^{(\mathbf{R})}(\mathbf{r}) = E_e^{(\mathbf{R})}\psi_e^{(\mathbf{R})}(\mathbf{r}) \quad (6.6)$$

where $E_e^{(\mathbf{R})}$ represents *the electronic energy when nuclei are set in configuration* $\mathbf{R}$. Putting the electron problem in this form implies that equation (6.6) must be separately solved for any possible nuclear configuration. This is an important conceptual difference with respect to atomic physics, where the problem of defining the electronic structure was unique: in molecular physics, instead, we have a multiplicity of ground and excited states, singularly referred to each specific nuclear arrangement. The corresponding eigenvalue problem for the roto-vibrational motion is

$$\left[-\frac{\hbar^2}{2}\sum_\alpha \frac{1}{M_\alpha}\nabla_\alpha^2 + \hat{V}_{nn}(\mathbf{R}) \right]\psi_n(\mathbf{R}) = [E_T - E_e^{(\mathbf{R})}]\psi_n(\mathbf{R}) \tag{6.7}$$

where the difference $E_n(\mathbf{R}) = [E_T - E_e^{(\mathbf{R})}]$ plays the role of molecular nuclear energy. This is a really important achievement: *molecular physics under the Born–Oppenheimer approximation basically consists in the separate solution of equation (6.6) and of equation (6.7)*.

In the present discussion such equations have been assumed to be the constitutive laws for the nuclear and the electronic problem, respectively, and this assumption has been in turn based on a reasonable guess. In appendix I their formal derivation is outlined, providing evidence that *the most fundamental fact supporting such a guess is that* $m_e/M_\alpha \ll 1$ for any possible nuclear mass value. This large difference of inertia translates into the fact that the electron motion can follow instantaneously the nuclear one: the time scales are so different that electrons feel nuclei as always clamped in their instantaneous configuration. This is in fact the physical condition under which we can separate the two problems. There is also another important consequence: since the electron and the nuclear systems are decoupled, *there is no way for them to exchange energy*. Under this respect, the Born–Oppenheimer approximation is also referred to as the *adiabatic approximation*.

For sake of completeness, we remark that non-adiabatic effects do exist [4] in molecules, although we will not treat them: in general, they are observed whenever nuclear motions promote couplings between different electronic states. The most important non-adiabatic occurrence is the so-called Jahn–Teller effect, resulting in a structural distortion of the molecular architecture in the attempt to minimize the electronic energy. It is even possible to observe the molecular pre-dissociation phenomenon: the molecule splits off in two (or more) parts as a consequence of an electronic transition to a continuum (unbound) state.

6.3 Molecular bonding

In order to understand the physics underlying the molecular bonding we will adopt the so-called *frozen core approximation*, according to which only valence electrons participate in the formation of the chemical bond. This feature has been extensively discussed in section 5.3 and allows one to adopt the following picture

$$\text{atom} = \underbrace{\text{nucleus} + \text{core electrons}}_{\text{ion}} + \text{valence electrons}$$

$$= \text{ion} + \text{valence electrons}$$

to describe an atom. An 'ion' is therefore a massive point-like object with positive charge. To keep the language as simple as possible, we will hereafter refer to 'valence electrons' just as 'electrons', since at this stage there is no possibility of misunderstanding: the core ones are frozen on the nucleus.

When two or more atoms give rise to a bound configuration (i.e. to a molecule), the wavefunction of each electron is largely affected by the nearby presence of positive ions and other negative electrons: this causes the lowering of the total energy of the system. We describe the net effect of these interactions by saying that *interatomic forces* are at work to form the molecule and, therefore, we understand that *the molecular bonding inherently has electromagnetic nature*.

There are basically two kinds of molecular bonding, namely the *ionic bond* and the *covalent bond*. In order to highlight their main qualitative features, we will treat them separately by discussing two prototypical case studies: the NaCl molecule and the H_2 molecule.

6.3.1 Ionic bonding: the NaCl molecule

Let us start with the NaCl molecule. The alkali Na atom belongs to the group IA (see figure 5.7) and, therefore, it contains just one $3s$ electron in the highest-energy incomplete electronic shell. Its ionization energy is $E_i^{Na} = 5.1$ eV: this is the work needed to generate a situation with a Na^+ ion and an isolated electron. The halogen Cl atom has, in turn, an electron affinity as large as $E_a^{Cl} = 3.8$ eV: this is the amount of energy released when an electron is added to its neutral ground state configuration. The energy balance is straight: in order to generate a couple of separated Na^+ and Cl^- ions we need to spend an energy of $E_i^{Na} - E_a^{Cl} = 1.3$ eV. However, it is plain to observe that two ions of opposite charge attract each other and, therefore, they tend to approach provided that they are free to move. Initially the total energy of the combined system steadily decreases by reducing the interatomic distance, but eventually the two ions could be so close that their electron clouds begin to overlap. This is not an energy-friendly situation, since it is at odds with the exclusion principle: some electron(s) must be necessarily promoted to higher energy state in order to fulfil the Pauli constraint. Furthermore, an electrostatic repulsion between the two closed shells and the two bare ions sets in. Overall, all these phenomena lead to a sharp increase in energy as the interatomic distance further decreases.

The situation so far outlined is summarized in figure 6.1, where the total potential energy $E_p(R)$ for NaCl molecule (in the gas phase) is shown as a function of the interatomic distance R and as the result of a non-trivial interplay among many different physical phenomena. A number of interesting information can be deduced from this picture. It is useful to proceed with a *gedankenexperiment*[1]: let us start with the two neutral atoms placed very far away and let us progressively reduce their distance. At first the interaction is vanishingly small and, therefore, the total potential energy of the two-atom system is constant (for convenience, this constant energy value is set equal to zero). However, at a separation of ~12–13 Å we observe

---

[1] This is a German word coined by H C Ørsted, meaning: thought experiment.

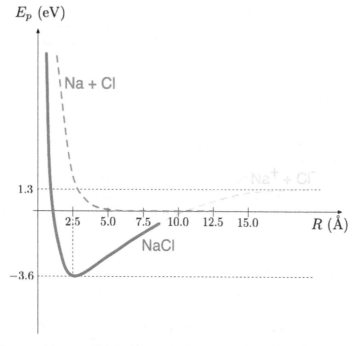

Figure 6.1. The potential energy $E_p(R)$ of a two-atom system made by Na and Cl as a function of the interatomic distance R. Red dashed line: the energy of the Na+Cl system. Cyan dashed line: the energy of the Na$^+$ and Cl$^-$ system. Full blue line: the minimum-energy curve of the bound system (molecule). The equilibrium distance $R_0 = 2.5$ Å and the minimum potential energy (dissociation energy) $E_p(R_0) = -3.6$ eV are shown by black dashed lines.

that some charge transfer from the alkali atom to the halogen one begins to occur. At shorter distances an increasingly strong Coulomb attraction sets in between the two ions with opposite charge, causing a dramatic reduction of the total potential energy. Its minimum value $E_p(R_0) \simeq -3.6$ eV is found at $R_0 = 2.5$ Å. This interatomic separation is named *equilibrium distance* of the molecule: here the bound system is lower in energy by 3.6 eV with respect to the two neutral atoms; such energy is commonly referred to as the *molecular dissociation energy* E_{diss}: it represents the work required to separate the molecule when it is initially placed in its state of minimum energy. It must be understood that the separation produces two neutral atoms[2]. By further reducing R, the combined effect of (i) Coulomb repulsion, and (ii) electron(s) promotion to a higher energy state caused by the Pauli principle offsets ionic attraction between Na$^+$ and Cl$^-$. This 'hard-wall' behaviour defines around each ion an exclusion volume which cannot be penetrated by the other chemical element.

---

[2] In other words: the dissociation energy is *not* the energy required to just separate at infinite distance the Na$^+$ and the Cl$^-$ ions (which, is shown in figure 6.1) is as large as $[1.3 - (-3.6)]$ eV $= 4.9$ eV but, rather, it is the work needed (i) to split the molecule, (ii) to recover the neutral atoms condition, and (iii) to separate them at infinite distance.

A widely used empirical expression of the energy $E_p(R)$ for a diatomic molecule is provided by the *Morse potential*

$$E_p(R) = E_{diss} \{1 - \exp[-\zeta_1(R - R_0)]\}^2 \tag{6.8}$$

where ζ_1 is an adjustable parameter typical of each molecule. At this stage we must consider it just as a phenomenological constant, whose actual physical nature will be discussed in the section 7.1.2 when molecular vibrations will be explicitly taken into account. This is in fact the main conceptual limitation of the Morse potential: it is only valid under the assumption of a rigid molecule, where no ionic motion takes place. In table 6.1 we report the equilibrium distance and dissociation energy values for some ionic diatomic molecules.

While the Morse potential can be used for covalently bonded molecules as well, *we can express the potential energy of a diatomic ionic molecule in a more specific form that highlights the electrostatic nature of the bonding* by using the following empirical expression

$$E_p(R) = -\frac{e^2}{4\pi\varepsilon_0}\frac{1}{R} + \frac{\zeta_2}{R^9} \tag{6.9}$$

where the inverse power law appearing in the second term has been set by fitting the experimental data, while the first term directly addresses the Coulomb attraction between the two ions. By imposing that at $R = R_0$ the energy $E_p(R)$ is minimum, we get either the expression of the ζ_2 parameter and of the dissociation energy, respectively

$$\zeta_2 = \frac{e^2 R_0^8}{36\pi\varepsilon_0} \quad \text{and} \quad E_{diss} = -\frac{8}{9}\frac{e^2}{4\pi\varepsilon_0}\frac{1}{R_0} \tag{6.10}$$

The accuracy limit of this empirical expression is determined, for instance, by comparing the experimental value of the dissociation energy for the NaCl molecule reported in table 6.1 with the corresponding $E_{diss} = 3.70$ eV value predicted by equation (6.10): the error is about 3% of the experimental datum, which is really good enough for many applications.

Table 6.1. Equilibrium distance R_0 and dissociation energy E_{diss} of some diatomic ionic molecules.

	R_0 (Å)	E_{diss} (eV)
HCl	1.27	4.43
LiH	1.60	2.50
NaCl	2.51	3.58
KF	2.55	5.90
KCl	2.79	4.92
KBr	2.94	3.96
KI	3.23	3.00
CsCl	3.06	3.76

Table 6.2. Electric dipole moment (measured in debyes) of some diatomic ionic molecules.

HCl	LiH	NaCl	KF	KCl	KBr	KI	CsCl
1.07	5.88	8.50	8.60	8.00	1.29	9.24	9.97

Because of their very nature, ionic molecules manifest a *permanent electric dipole moment*: for instance, if we consider the NaCl molecule at its equilibrium distance, we can describe it as a pair of $\pm e$ charges placed at distance R_0. This results in a net electric dipole moment as large as 8.5 debyes[3]. In table 6.2 we report the dipole moments of some diatomic ionic molecules: the way dipoles orientate under the action of an external electric field determines the dielectric properties of molecular liquids or crystals [5]. Ionic molecules are referred to as *polar molecules*, characterized by *heteropolar bonding*.

6.3.2 Covalent bonding: the H_2 molecule

Covalent bonding can be understood only quantum mechanically, at variance with the ionic case which can be ultimately be described in terms of classical Coulomb interactions. This makes everything much more intriguing, although more subtle. At this stage we have not yet developed any quantum theory of the molecular electronic structure and, therefore, we will proceed by qualitative reasoning, the theoretical basis for which will be found in chapter 8. The case study is the H_2 molecule which we will approach in steps, starting to consider its simpler ionic configuration: the hydrogen molecular ion H_2^+.

The H_2^+ molecule consists in two protons and just one electron. It has axial symmetry or, equivalently, its physical properties are invariant upon rotation around the direction connecting the two H nuclei, hereafter referred to as the *molecular axis*. Since it is homonuclear, its properties are also symmetrical with respect to a plane normal to the molecular axis and passing through the intermediate point between the two nuclei, hereafter referred to as the *reflection plane*. Because of Coulomb coupling, both nuclei attract the lone electron, while repelling each other: the balance of these two phenomena (whose relative strength depends on the internuclear distance R) provides the total potential energy $E_p(R)$.

Thanks to the molecular symmetry, *the probability of finding the electron in two points A and B placed at the same distance on opposite sides of the reflection plane must be the same*. If we suppose to know the electron wavefunction ψ, then we can set $\psi^*(A)\psi(A) = \psi^*(B)\psi(B)$. We are of course working under the Born–Oppenheimer approximation and we are considering only the space part of the wavefunction (the electron spin will come into play soon). Mathematics dictates that this condition is only satisfied provided that either $\psi(A) = \psi(B)$ or $\psi(A) = -\psi(B)$: in other words, *the electron wavefunction can only be either even or odd with respect to the reflection plane*. The main physical consequence of this fact is that the odd

---

[3] In the international system, the unit of electric dipole moment is set as: 1 debye = 3.3×10^{-30} m C.

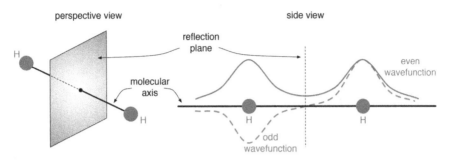

Figure 6.2. Schematic illustration of the even or odd character of the electron wavefunction in the H_2^+ molecular ion.

wavefunction is constrained to be zero at the molecule centre, while the even wavefunction is not required to comply with this imposition. The situation is schematically depicted in figure 6.2 where the ψ is evaluated along the molecular axis.

The *odd* and the *even electron*[4] behave quite differently: the first one tends to avoid the central molecular region[5], thus paving the way for a strong nuclear repulsion. In addition, a low probability density at the centre determines a much greater probability to find such an electron outside the nuclear regions[6], where it is only loosely bound. In summary: whenever the electron of a H_2^+ molecule is accommodated on a state described by an odd wavefunction, the molecule is very weakly (or not at all) bound. In chapter 8 we will name such a state an *anti-bonding state*. In contrast, the even electron has a comparatively much higher probability of being in the central region, where it is tightly bound and, furthermore, it efficiently screens the nuclear Coulomb repulsion. The combination of these two facts favours the formation of a stable bonded molecular configuration, usually referred to as a *bonding state*.

The discussion so far developed has implicitly assumed that the internuclear distance was fixed at a given value R. Accordingly, we could only compare the energy of an anti-bonding state with that of a bonding one, at the same value of R. If we now repeat the same *gedankenexperiment* described in section 6.3.1 we obtain the result shown in figure 6.3, which provides robust evidence of *a qualitatively different behaviour between bonding and anti-bonding states*. In the former case, since the negative (attractive) electron–nucleus energy dominates over the positive (repulsive) nucleus–nucleus energy, the resulting $E_p(R)$ lowers as the internuclear distance decreases, until reaching the absolute minimum at the equilibrium distance $R_0 = 1.06$ Å. By bringing the two nuclei even closer, the nuclear repulsion eventually overcomes the electron-mediated attractive energy and we enter in the hard-wall regime. In contrast, the $E_p(R)$ for the anti-bonding state is monotonically increasing

---

[4] These are a useful abbreviations instead of: 'the electron described by an odd/even wavefunction'.
[5] Simply because its presence probability density is there very small, and exactly zero at the molecule centre.
[6] This is of course due to the normalization of the electron wavefunction.

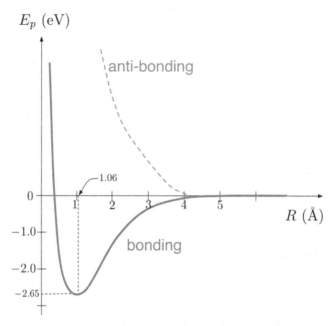

Figure 6.3. The potential energy $E_p(R)$ of the H_2^+ molecular ion in a bonding (blue full line) and anti-bonding (red dashed line) state. Its minimum $E_p(R_0) = -2.65$ eV at $R_0 = 1.06$ Å is shown by a black dashed line.

as R decreases, since positive repulsion always exceeds negative attraction. This implies that there is no minimum, i.e., no stable bound configuration.

Everything we did so far was just in preparation for the discussion of H_2 molecule and, therefore, we now ideally add the second electron to the hydrogen molecular ion. Its wavefunction obeys the same symmetry properties previously discussed for the first electron. This implies that if we are considering the molecular ground state (which, of course, is a bound state), *the two space wavefunctions are identical and have the same even symmetry*. This is a key feature imposing on us that we cannot any longer neglect the role of spin. In order to keep our arguments at a conceptually rigorous but formally elementary level, we will treat spin similarly to the discussion developed in section 5.3, when we were dealing with the *aufbau*. Accordingly, the two electrons of the H_2 molecule in its ground state must have opposite spin[7]. Another new issue is imposed by the fact that we have a system of two indistinguishable electrons with partially overlapping wavefunctions. This means that *we cannot associate any of them with a specific nucleus*, a conclusion which is usually summarized by the following statement: *the two electrons are shared by the two atoms forming the molecule*. The emerging picture is that in order to form the bonding condition for the H_2 molecule we need a twofold occurrence: (i) two electrons must be shared, with (ii) opposite spin. When this eventuality is found, we say that *a covalent bond* has been formed.

---

[7] This guarantees that, against a symmetrical space part, the spin part is antisymmetric, so that the total wavefunction is the correct one for the system of two electrons.

Table 6.3. Equilibrium distance R_0 and dissociation energy E_{diss} of some diatomic covalent molecules, both homo- and heteronuclear.

	R_0 (Å)	E_{diss} (eV)
H_2	0.74	4.48
Li_2	2.67	1.03
N_2	1.09	7.37
O_2	1.21	5.08
Cl_2	1.99	2.47
HI	1.61	3.06
CO	1.13	11.11
NO	1.15	5.30

From our discussion it is straightforward to draw another important conclusion: *no more than two electrons can participate in the same covalent bond*. An atom having more than one electron in the highest-energy incomplete sub-shell could in principle form as many covalent bonds with nearby atoms. This holds, however, only for unpaired electrons[8]. In different words: pairs of valence electrons with antiparallel spin cannot participate in an interatomic covalent bond. This implies that, for instance, nitrogen could form up to three covalent bonds, while this possibility is limited to two bonds for oxygen (see figure 5.6).

In conclusion, we remark that *covalent bonding is inherently directional*, as has clearly emerged from our discussion of the simple H_2 case study. Also, *it can take place even between two identical atoms*, in which case the molecule is said to be *homonuclear*. Both characteristics are not found in ionic bonding, which is spherically symmetrical and always heteronuclear. Finally, the Morse potential given in equation (6.8) provides a reasonably good (empirical) estimation of the covalent bonding energy. In table 6.3 we report equilibrium distances and dissociation energies for some selected diatomic covalent molecules.

Further reading and references

[1] Bransden B H and Joachain C J 1983 *Physics of Atoms and Molecules* (Harlow: Addison-Wesley Longman)

[2] Demtröder W 2010 *Atoms, Molecules and Photons* (Heidelberg: Springer)

[3] Eisberg R and Resnick R 1985 *Quantum Physics of Atoms, Molecules, Solids, Nuclei, and Particles* 2nd edn (Hoboken, NJ: Wiley)

[4] Atkins P and Friedman R 2011 *Molecular Quantum Mechanics* 5th edn (Oxford: Oxford University Press)

[5] Hook J R and Hall H E 2010 *Solid State Physics* (Hoboken, NJ: Wiley)

---

[8] According to section 5.3 'unpaired' electrons have parallel spins.

IOP Publishing

Atomic and Molecular Physics (Second Edition)
A primer
Luciano Colombo

Chapter 7

Molecular vibrations and rotations

Syllabus—*In the spirit of the Born–Oppenheimer approximation presented in chapter 6, we will discuss the nuclear motions in a molecule lying in an assigned electronic configuration. First, we will discuss the case of diatomic molecules, whose rotational and vibrational spectra will be calculated with great accuracy, including the roto-vibrational coupling. Next, we will discuss the interaction between a radiation bath and the nuclear degrees of freedom, considering both the photon emission/absorption and the photon scattering (Raman and Rayleigh effects) mechanisms. Finally, we will extend our investigation to polyatomic molecules, providing a semi-quantitative introduction to their rotations and vibrations. To this aim we will outline some more advanced approaches, based on molecular symmetry properties and normal coordinates.*

7.1 Molecular motions in diatomic molecules

In principle a molecule can *diffuse*, *rotate*, and *vibrate*. The first kind of motion is conveniently described as the diffusion of a rigid body, whose coordinates coincide with those of the molecular centre-of-mass, and typically treated by classical kinetic theory. In fact, we are not interested in this physics, since *we want to focus on the inner molecular motions which are fully described by equation (6.7)*. Under the Born–Oppenheimer approximation, this equation accounts for both rotational and vibrational motions, which are experimentally found to have typical frequency in the range 10^{11}–10^{12} Hz and 10^{13}–10^{14} Hz, respectively. Therefore, following a procedure already adopted many times whenever the complete problem can be read as the superposition of phenomena with unlike energy scales, *we will attempt to separate rotations from vibrations* and then we will individually investigate the corresponding physics. The procedure to obtain this goal is really straightforward if we focus on diatomic molecules: here the formalism is simple and physically transparent and, more importantly, we can take benefit from some previous achievements.

Let the two atoms forming the molecule have mass M_1 and M_2 and position $\mathbf{R}_1$ and $\mathbf{R}_2$, respectively; then, we define the *molecular reduced mass* $M = M_1M_2/(M_1 + M_2)$ and the *relative position vector* $\mathbf{R} = \mathbf{R}_1 - \mathbf{R}_2$. The reduced nuclear problem[1] in a diatomic molecule can be treated formally as the hydrogen atom (see section 3.1.1): by representing the relative position vector by its polar coordinates $\mathbf{R} = (R, \theta, \phi)$, we can write the nuclear Hamiltonian as

$$\hat{H}_{\text{nucl}} = \left[-\frac{\hbar^2}{2M} \frac{1}{R^2} \frac{\partial}{\partial R}\left(R^2 \frac{\partial}{\partial R}\right) + \frac{1}{2MR^2} \hat{L}_{\text{rot}}^2 + \hat{V}_{\text{nucl}} \right] \tag{7.1}$$

where $\hat{L}_{\text{rot}}^2$ is the square angular momentum operator describing the rotational motion. On the other hand, $\hat{V}_{\text{nucl}}$ is the potential energy operator corresponding to the classical expression

$$V_{\text{nucl}}(\mathbf{R}) = V_{\text{nn}}(\mathbf{R}) + E_e^{(\mathbf{R})} \tag{7.2}$$

for the nuclear degrees of freedom; it includes the so-called *adiabatic contribution* $E_e^{(\mathbf{R})}$ associated with the molecular electronic state, as we learned in the previous section.

Equation (7.1) is the nuclear counterpart of the Hamiltonian operator defined in equation (3.8) for the electron of the hydrogen atom and, therefore, the corresponding eigenvalue equation

$$\hat{H}_{\text{nucl}}\psi_n = E_T\psi_n \tag{7.3}$$

can be formally solved in a similar way: we are allowed to seek for a solution

$$\psi_n = \psi_{\text{rot}}(\theta, \phi)\psi_{\text{vib}}(R) \tag{7.4}$$

where we introduced the *rotational* $\psi_{\text{rot}}(\theta, \phi)$ and the *vibrational* $\psi_{\text{vib}}(R)$ wavefunctions.

Our previous knowledge helps us to easily understand that the rotational wavefunction is a spherical harmonic function

$$\psi_{\text{rot}}(\theta, \phi) = Y_{rm_r}(\theta, \phi) \tag{7.5}$$

where *r is the rotational quantum number* providing the eigenvalues of the square rotational angular momentum

$$\hat{L}_{\text{rot}}^2 \, Y_{rm_r}(\theta, \phi) = r(r + 1)\hbar^2 \, Y_{rm_r}(\theta, \phi) \tag{7.6}$$

while m_r is the quantum number providing its z-component $\hat{L}_{\text{rot},z}$

$$\hat{L}_{\text{rot},z} \, Y_{rm_r}(\theta, \phi) = m_r\hbar Y_{rm_r}(\theta, \phi) \tag{7.7}$$

By inserting equation (7.6) into equations (7.1) and (7.3) and by setting $\psi_{\text{vib}}(R) = u_r(R)/R$ as explained in appendix A, we get

---

[1] This is an abbreviation to mean: the rotation and the oscillation of the effective mass around the molecular centre-of-mass.

$$-\frac{\hbar^2}{2M}\frac{d^2u_r(R)}{dR^2} + \left[\hat{V}_{\text{nucl}} + \frac{r(r+1)\hbar^2}{2MR^2}\right]u_r(R) = E_T u_r(R) \tag{7.8}$$

which represents the eigenvalue equation for the vibrational motions. Interestingly enough, *the term $r(r+1)\hbar^2/2MR^2$, appearing in this equation makes vibrational states depend upon the rotational ones*. It is usually referred to as the 'centrifugal term'. We must duly take note of this: *a complete theory should include such roto-vibrational coupling*.

In order to fine-tune the hierarchy of interdependencies existing among nuclear motions, it is appropriate at this stage to introduce one further approximation. It phenomenologically results that vibrational amplitudes in a diatomic molecule are small with respect to the equilibrium distance R_0. It is therefore useful to introduce the *relative displacement variable* $(R - R_0)$ and to expand both terms in the square parenthesis of equation (7.8) only up to the first order. More rigorously: we expand their classical expression, which is next translated into a new operator. For the V_{nn} term appearing in V_{nucl} (see equation (7.2)) we get the following approximated expression

$$V_{\text{nn}}(\mathbf{R}) \simeq \frac{1}{2}\frac{d^2V_{\text{nn}}}{dR^2}\bigg|_{R_0}(R - R_0)^2 = \frac{1}{2}\gamma\,(R - R_0)^2 \tag{7.9}$$

where we have set $(d^2V/dR^2)_{R_0} = \gamma$. Similarly, for the second term we get

$$\frac{1}{2MR^2}\hat{L}_{\text{rot}}^2 = \frac{r(r+1)\hbar^2}{2M}\frac{1}{R^2} \simeq \frac{r(r+1)\hbar^2}{2M}\frac{1}{R_0^2} \tag{7.10}$$

so that eventually we have

$$V_{\text{nucl}}(\mathbf{R}) = \frac{1}{2}\gamma\,(R - R_0)^2 + E_e^{(\mathbf{R_0})} \tag{7.11}$$

from which is straightforwardly obtained the corresponding $\hat{V}_{\text{nucl}}(\mathbf{R})$ operator. Equation (7.9) suggests that the two atoms forming the molecule interact through a parabolic potential: this picture is called *harmonic approximation* and rules the nuclear oscillations around some interatomic distance. In this regard, we observe that such a distance is in fact slightly different from the static equilibrium one R_0, because of the stretching effect due to the rotations described by equation (7.10).

The picture so far developed defines a multiple-step approach to the study of nuclear motions [1–4]. In the first instance, *molecular rotations are studied within the 'rigid molecule' approximation* (where the interatomic distance is set at R_0) provided by equation (7.10); this will also allow the complete decoupling between vibrations and rotations, since the centrifugal term in equation (7.8) is now reduced to a constant. Next, *molecular vibrations are studied in the harmonic approximation* as oscillations taking place around a suitably defined equilibrium distance[2]. Eventually,

---

[2] In other words, rotational stretching could be neglected or included, according to the level of sophistication we are interested in.

roto-vibrational coupling and anharmonic affects are duly taken into account by a perturbative approach. In the following three sections we will proceed as indicated, since it is the more physically sound approach.

7.1.1 Rotational spectra

A rigid diatomic molecule with effective mass M is classically described by a rotational energy $E_{rot} = L_{rot}^2/2I$, where $I = MR_0^2$ is its moment of inertia for rotations about the direction normal to the molecular axis and passing through the centre-of-mass[3]. Thanks to equation (7.6) this translates into the quantum expression

$$E_{rot} = \frac{\hbar^2}{2MR_0^2} r(r+1) = \frac{\hbar^2}{2I} r(r+1) = Bhc \; r(r+1) \tag{7.12}$$

where we have introduced the abbreviation

$$B = \frac{\hbar}{4\pi MR_0^2 c} = \frac{\hbar}{4\pi Ic} \tag{7.13}$$

defining the so-called *rotational constant* of the molecule. Equation (7.12) states that *the rotational spectrum of a diatomic molecule is discrete*. The separation between the two successive rotational levels r and $(r+1)$ is given by the following rule

$$\Delta E_{rot} = 2 \; Bhc \; (r+1) \tag{7.14}$$

and this generates the energy diagram shown in figure 7.1 (left). The values of the $\hbar^2/2I$ ratio[4] of some homo- and heteropolar diatomic molecules are shown in table 7.1.

We observe that the rotational constant typically falls in the range 10^{-5}–10^{-3} eV, thus ending up always smaller than the energy of the room temperature thermal bath $k_B T_{room} \sim 25 \times 10^{-3}$ eV. Therefore, in a gas at room temperature most of its molecules are likely promoted to excited rotational levels[5]. We can evaluate by a simple argument the population of the rotational levels in a gas sample made by diatomic molecules in equilibrium at temperature T. From the Boltzmann statistics (see appendix E) we calculate the number N_r of molecules in the rth rotational level as

$$N_r = \frac{N}{Z_{rot}} g_r \exp(-E_{rot}/k_B T) \tag{7.15}$$

---

[3] Obviously, the moment of inertia about the molecular axis is negligibly small since the electron mass is much smaller than any ionic one.

[4] We remark that in some textbooks the $\hbar^2/2I$ ratio is actually referred to as the rotational constant of the molecule. Here for such a constant we rather adopt the definition given in equation (7.13).

[5] The small value of the rotational constant has also another important consequence: the direct experimental determination of the pure rotational spectrum can only be obtained in a molecular gas at very low pressure, otherwise collisions likely make the discrete spectrum characterized by very narrow energy levels transform into a continuum one.

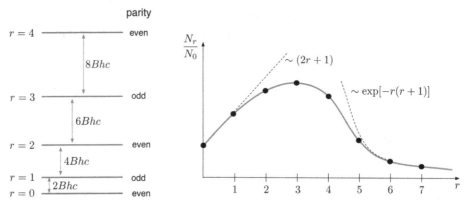

Figure 7.1. Left: the rotational energy diagram of a generic diatomic molecule. The parity of the rotational wavefunction is indicated and the allowed electric dipole transitions (both in emission and absorption) are accordingly shown by arrowed vertical lines. Right: the population of the different rotational levels in a molecular gas at finite temperature. Since r is a discrete variable, the full red line must be intended just as a guide to the eye.

Table 7.1. The $\hbar^2/2I$ ratio (eV units) for some diatomic molecules.

	H_2	Li_2	N_2	O_2	Cl_2
$\hbar^2/2I$	7.56×10^{-3}	8.39×10^{-5}	2.48×10^{-4}	1.78×10^{-4}	3.10×10^{-5}
	HCl	NaCl	KCl	KBr	CO
$\hbar^2/2I$	1.32×10^{-3}	2.36×10^{-5}	1.43×10^{-5}	9.10×10^{-6}	2.38×10^{-4}

where N is the number of molecules, $\mathcal{Z}_{\text{rot}}$ is the rotational partition function [5, 6], which for the present derivation we do not need to further specify, and $g_r = 2r + 1$ is the degeneracy of the level, reflecting the possible different values of the quantum number $m_r = 0, \pm 1, \pm 2, \ldots, \pm r$. By inserting equation (7.12) we straightforwardly get

$$\frac{N_r}{N_0} = (2r + 1) \exp[-Bhc\ r(r + 1)/k_B T] \tag{7.16}$$

providing the ratio between the number of molecules in the rth and in the fundamental ($r = 0$) rotational levels when the system is in equilibrium at temperature T; the result is shown in figure 7.1 (right). The balance between the linear ($2r + 1$) increase and the exponential decay determines a non-monotonic trend, showing a maximum which falls in between $r = 2$ and $r = 4$ for most of the molecules reported in table 7.1.

If the diatomic molecule is heteropolar, then it has a permanent electric dipole moment, as discussed in section 6.3.1. *The molecular dipole can interact with an electromagnetic wave, giving rise to transitions occurring between rotational levels*[6]. They typically emit or absorb photons in the far-IR or MW region and must fulfil selection rules. We know from equation (7.6) that rotational wavefunctions are spherical harmonic functions and, therefore, they have $(-1)^r$ parity. This implies that electric dipole transitions, similarly to the case of hydrogenic atoms, can only occur between states with opposite parity. In other words, *for the rotational spectrum it holds the selection rule* $\Delta r = \pm 1$. The inverse spectral wavelength of emitted/absorbed photons is easily calculated from equation (7.14)

$$\lambda^{-1} = \frac{\hbar}{2\pi Ic} (r + 1) = 2B (r + 1) \qquad (7.17)$$

where $(r + 1)$ is the rotational quantum number of the highest state, as intended before. Since it must hold that $\Delta r = \pm 1$, then

$$\lambda_{r+1}^{-1} - \lambda_r^{-1} = \Delta \lambda^{-1} = \frac{\hbar}{2\pi Ic} = 2B \qquad (7.18)$$

or, equivalently, *the separation* $\Delta \lambda^{-1}$ *between the lines of the rotational spectrum is constant*. This is a very interesting and useful result: a direct spectroscopic measure of $\Delta \lambda^{-1}$ provides the moment of inertia and so the equilibrium interatomic distance of the molecule. A typical rotational spectrum appears as in figure 7.2: the intensity of each line reflects the population of the various rotational levels.

We conclude this section with two remarks. We developed our discussion under the approximation of rigid molecule. If we now admit that by increasing the

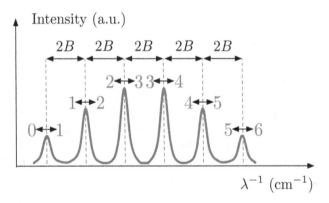

Figure 7.2. A schematic representation of the emission/absorption intensity in a typical rotational spectrum of a polar diatomic molecule with rotational constant B. The rotational quantum numbers involved in the transition $r \leftrightarrow (r + 1)$ (with $r = 0, 1, 2,...$) are indicated for each spectral line. It is supposed that N_r is maximum for $r = 3$.

---

[6] We remark that homonuclear diatomic molecules (i.e., molecules with no permanent dipole moment) can undergo pure rotational transitions only if an electronic transition simultaneously takes place with the rotational one. This process is in general less likely than the pure rotational transition occurring in heteropolar molecules.

rotational energy a stretching of the interatomic distance is observed, then we must conclude that the molecular moment of inertia is correspondingly affected, because of the centrifugal effect: a correction should be added to equation (7.12). Finally, we remark that the results of this section also apply to multi-atomic molecules, provided that they are linear like, e.g., the CO_2 molecule; the main difference is the value of the moment of inertia which is of course larger and, therefore, rotational levels are closer and lower in energy than in diatomic molecules.

7.1.2 Vibrational spectra

Let us now consider the vibrational problem under the harmonic approximation. This allows us to rewrite equation (7.8) in the simplified form

$$-\frac{\hbar^2}{2M}\frac{d^2u(x)}{dx^2} + \frac{1}{2}\gamma\, x^2\, u(x) = E_{\text{vib}}\, u(x) \tag{7.19}$$

where we have omitted the label r in the symbol of the eigenfunction $u(x)$ in order to keep the formalism simple (this issue will be readdressed later), we have indicated by x the variation of the equilibrium distance (at this stage it is irrelevant whether we choose R_0 or the rotationally-stretched one), and we have introduced the *pure vibrational energy* $E_{\text{vib}} = E_T - E_e^{(R_0)} - E_{\text{rot}}$ of the molecule (where E_{rot} is calculated according to equation (7.12)).

Equation (7.19) states that the vibrational physics of a diatomic molecule corresponds to a *one-dimensional quantum harmonic oscillator*. The formal solution is outlined in appendix J, where eigenfunctions are proved to be Hermite polynomials and energies are found to be quantized as

$$E_{\text{vib}} = \left(n + \frac{1}{2}\right)\hbar\omega_0 \ \text{ with } n = 0, 1, 2, 3,... \tag{7.20}$$

where n is the *vibrational quantum number* and $\omega_0 = \sqrt{\gamma/M}$ is the fundamental oscillation frequency, which is reported in table 7.2 for some selected cases.

The vibrational spectrum of a diatomic molecule described by equation (7.20) is discrete and non-degenerate. Furthermore, its levels are equidistant: it is easy to calculate their separation $\Delta E_{\text{vib}} = \hbar\omega_0$. This implies that, provided it has a permanent electric dipole moment, a diatomic molecule can emit or absorb photons of energy $E_{\text{photon}} = \hbar\omega_0$, which typically correspond to electromagnetic waves in the IR region.

Table 7.2. Fundamental vibrational energy (eV units) of some diatomic molecules.

	H_2	Li_2	N_2	O_2	Cl_2
$\hbar\omega_0$	0.543	0.0434	0.292	0.194	0.0698
	HCl	NaCl	KCl	KBr	CO
$\hbar\omega_0$	0.369	0.047	0.034	0.028	0.268

Similarly to the case of optical spectra of any other kind, the transitions between vibrational levels must fulfil selection rules dictated by the parity of their corresponding eigenfunctions. In appendix J it is demonstrated that the Hermite polynomial of nth order has parity $(-1)^n$ and, therefore, *vibrational transitions obey the selection rule* $\Delta n = \pm 1$. The situation is summarized in figure 7.3.

We finally readdress the discussion of the Morse potential introduced in section 6.3.1. Under the adopted assumption of small oscillations, we can expand the exponential factor in powers of $(R - R_0)^2$ and retain just the first two leading terms, so that the Morse potential can be approximated to

$$E_p(R) \simeq E_{diss}\, \zeta_1^2\, (R - R_0)^2 \tag{7.21}$$

In this harmonic approximation we also have

$$E_p(R) = \frac{1}{2}\, \gamma\, (R - R_0)^2 \tag{7.22}$$

and therefore, by comparing the two above forms for the same $E_p(R)$ energy, we easily obtain a microscopic definition of the ζ_1 constant, so far treated as a phenomenological parameter. We get

$$\zeta_1 = \sqrt{\frac{M}{2E_{diss}}}\; \omega_0 \tag{7.23}$$

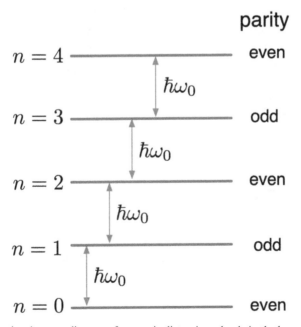

Figure 7.3. The vibrational energy diagram of a generic diatomic molecule in the harmonic approximation. Purely vibrational emission/absorption transitions are indicated by arrowed vertical lines. The vibrational quantum number v (left), the parity of the vibrational eigenfunctions (right), and the separation $\hbar\omega_0$ (corresponding to the energy of the emitted/absorbed photon) are also indicated.

a result which provides ζ_1 in terms of well-known molecular properties.

In conclusion, we remark that *vibrational spectra provide direct information about the relative abundance of natural isotopes.* Since the energy $\hbar\omega_0$ of the emitted or absorbed IR photons does depend on the molecular reduced mass, any isotope is characterized by its own emission/absorption line. By measuring their intensities, we easily get their relative abundance.

7.1.3 Roto-vibrational spectra

The total roto-vibrational energy $E_{\text{rot-vib}}$ of a diatomic molecule is straightforwardly written by combining equation (7.12) with equation (7.20)

$$E_{\text{rot-vib}} = E_{\text{rot}} + E_{\text{vib}} = Bhc\ r(r+1) + \left(n + \frac{1}{2}\right)\hbar\omega_0 = E_{\text{T}} - E_{\text{e}}^{(\text{R})} \qquad (7.24)$$

where the last equality has been added to remind that, in the spirit of the Born–Oppenheimer approximation, *this amount of energy corresponds to the total molecular energy from which has been subtracted the electronic contribution.*

By comparing tables 7.1 and 7.2 we immediately realise that, in general, for any given molecule the rotational constant is much smaller than the fundamental vibrational energy. This implies that *several rotational levels correspond to each vibrational one* as shown by the *roto-vibrational energy diagram* reported figure 7.4 (left). It must be noted that only rotational levels with $r > 1$ are separated in energy with respect to their parental vibrational level.

Since the true energy diagram associated with nuclear motions is inherently roto-vibrational, *it is possible to have transitions between rotational levels belonging to different vibrational ones.* Such transitions, however, must obey both $\Delta r = \pm 1$ and

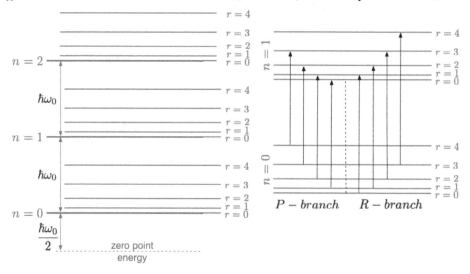

Figure 7.4. Left: the roto-vibrational energy diagram of a generic diatomic molecule in the harmonic approximation. Right: some allowed absorption transitions between roto-vibrational levels. The central dashed line corresponds to the forbidden $\Delta r = 0$ transition. The $P - branch$ and the $R - branch$ correspond, respectively, to $\Delta r = -1$ and $\Delta r = +1$ transitions.

$\Delta n = \pm 1$ selections rules. This implies that roto-vibrational transitions can only occur between two rotational levels (i) with consecutive rotational quantum numbers and (ii) belonging to adjacent vibrational levels.

Roto-vibrational absorption transitions from the state $n = 1$ to the state $n = 0$ are illustrated in figure 7.4 (right), from which a number of intriguing features emerge. First of all, we note that the spectral lines form two series, symmetrically centred at $\bar{\nu} = \omega_0/2\pi$. Their frequency is given by the following equation

$$\nu = \bar{\nu} \pm 2Bc\,(r + 1) \tag{7.25}$$

where $(r + 1)$ is the upper rotational level interested by the transition. This result indicates that the transition frequencies are equally spaced by the amount $2Bc$. Finally, it must be noted that the line at frequency $\bar{\nu}$ is missing because it corresponds to a forbidden transition with $\Delta r = 0$. The lines for which $\Delta r = +1$ or $\Delta r = -1$ are said to form the $R - Branch$ or the $P - Branch$, respectively.

High-precision roto-vibrational spectroscopic measurements provide direct information about the unalike centrifugal stretching occurring at each vibrational level (making the line separation not actually constant), also about the anharmonicity effects neglected in the expansion given by equation (7.9) [1, 3, 4].

7.2 Rayleigh and Raman scattering

Besides spectroscopy, *it is possible to study a molecular system also by means of light scattering experiments*: the focus is now on *how photons are scattered by a molecule*, rather than on how they are emitted or absorbed.

Light scattering is a second-order process [1, 2] that, at the level we are developing our treatment, can be simply described by the following sequence of events: an incoming photon is absorbed by a molecule and a second photon is eventually re-emitted by molecular deexcitation. The experimental implementation of a light scattering experiment is conceptually simple: a molecular gas is illuminated with a monochromatic radiation and the scattered light is observed at proper angles. If the incoming photons have energy $\hbar\omega_{\mathrm{in}}$, then it is typically observed that *scattered photons have either the very same energy or they have energy $\hbar\omega_{\mathrm{in}} \pm \hbar\bar{\omega}$*, where $\hbar\bar{\omega}$ corresponds to a quantum of molecular energy. Emission of photons at the same frequency ω_{in} of the illuminating radiation is called *Rayleigh scattering*, while the emission at frequency $\omega_{\mathrm{in}} \pm \bar{\omega}$ is called *Raman scattering*.

We can elaborate a simple phenomenological picture for any kind of scattering phenomena by arguing that, following the initial absorption, the molecule is excited to a virtual (i.e., not stationary) unstable state. Therefore, it immediately undergoes a re-emission transition, by reaching either the same initial state (Rayleigh scattering) or a different one (Raman scattering). In this latter case, we must further distinguish between two different situations: if the molecule was initially in its ground state, then the final state reached after Raman scattering must be higher in energy; on the other hand, if the molecule was initially in any excited state, then the final one reached by re-emission could be higher or lower in energy. Deexcitation to an higher-energy level implies that the radiation is re-emitted at frequency $\omega_{\mathrm{in}} - \bar{\omega}$ generating a so-called *Stokes line*. The opposite case of radiation re-emitted at frequency $\omega_{\mathrm{in}} + \bar{\omega}$ generates an *anti-Stokes* line.

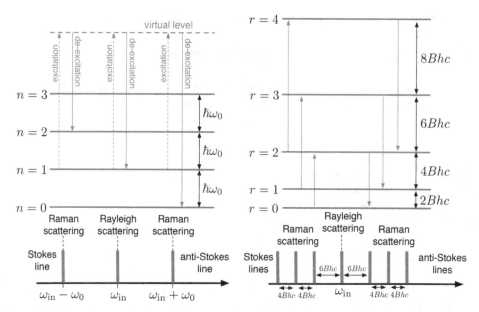

Figure 7.5. Schematic representation of the Raman scattering process, transitions, and corresponding spectra. Two different illustration schemes have been used for transitions between vibrational (left) and rotational (right) states. See text for a detailed explanation.

Raman scattering (see figure 7.5) can occur between either rotational and vibrational levels and, in any case, must fulfil selection rules. *The vibrational Raman scattering is governed by the selection rule* $\Delta n = \pm 1$, since it is proved that it is formally equivalent to ordinary purely vibrational transitions [4]. On the other hand, *the rotational Raman scattering must obey the selection rule* $\Delta r = \pm 2$, which is easily explained in the following terms: both the first excitation transition and the following deexcitation one must fulfil the ordinary $\Delta r = \pm 1$ rule, so that the total change in the quantum rotational number could be either 0 (corresponding to the Rayleigh scattering) or ± 2. In conclusion, *rotational Raman transitions can occur at frequencies that are forbidden in ordinary emission/absorption rotational spectroscopy.*

In figure 7.5 (left) is shown an excitation/deexcitation transition to/from the intermediate virtual state in the case of a vibrational Raman event for both the Stokes and the anti-Stokes process; the Rayleigh virtual transition is shown as well. These transitions generate two Raman lines and one Rayleigh line as shown in the bottom. In figure 7.5 (right), instead, some Stokes and some anti-Stokes transitions are represented directly on the energy diagram in the case of rotational Raman events. In this case the excitation/deexcitation to/from the virtual level is not shown and, therefore, arrowed lines directly link the initial and the final states interested by the scattering event[7]. On the bottom the corresponding spectrum is reported, with

---

[7] In other words: they do not indicate a direct emission/absorption transition as in the case of pure vibrational or rotational spectroscopy.

indication of the energy separation between lines. The two parts of this figure illustrate the typical ways of graphically representing the Raman effect.

We conclude this section by an important remark: the Raman effect does not require the existence of any molecular permanent electric dipole moment and, therefore, *Raman lines are observed even in homonuclear molecules* which have no vibrational or rotational absorption/emission spectra. In this respect, Raman measurements are a very useful complementary technique to spectroscopy.

7.3 Nuclear motions in polyatomic molecules

The quantitative description of rotations and vibrations in polyatomic molecules is a rather complex topic, which is rigorously treated elsewhere [2, 3]. Here we only provide a brief summary, based on three basic notions developed in the diatomic case, namely: (i) we will assume that rotations and vibrations can be separately treated, (ii) we will make use of the rigid molecule approximation for rotations, and (iii) we will describe vibrations within the harmonic approximation.

7.3.1 Rotations

At variance with the diatomic case, *a polyatomic molecule can rotate around any axis passing through its centre-of-mass* and its total angular momentum **L** is constant in time, provided that the molecule is not under the action of any external torque. In general, therefore, *a free polyatomic molecule rotates around an axis which, in turn, undergoes nutation around the fixed direction of the angular momentum.*

The starting point to describe such a complex rotational motion is the classical energy for a rigid rotator [7]

$$E_{rr} = \frac{L_a^2}{2I_a} + \frac{L_b^2}{2I_b} + \frac{L_c^2}{2I_c} \tag{7.26}$$

where I_a, I_b, and I_c are the principal moments of inertia, while L_a, L_b, and L_c are the corresponding components of the angular momentum **L** with respect to the principal axes $\{a, b, c\}$. The space orientation of such axes is determined by the symmetry properties of the molecule, which also impose relationships among the three moments of inertia.

Because of the major role played by symmetry, the general solution of the rotational problem requires the use of group theory [4], thus placing this topic beyond the scope of this manual. However, at least the special case of molecules with a rotational symmetry axis can be treated with elementary considerations since in this case the molecule is formally equivalent to a symmetric rotor for which $I_a \neq I_b = I_c$, where it has been assumed that the symmetry axis is the a principal axis. Such molecules are referred to as *symmetric top molecules*. Relevant examples are provided by linear molecules (like e.g., CO_2, NCO or HCN), by planar molecules with a the symmetry axis normal to the molecular plane (like e.g., SO_3, and C_6H_6) and by polyatomic molecules with an axial symmetry around a bonding direction (like e.g., CH_3I and CH_3F). Some examples are shown in figure 7.6. For a symmetric rotor equation (7.26) takes the simple form

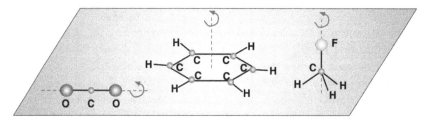

Figure 7.6. Examples of symmetric top molecules with a rotational symmetry axis (blue dashed line).

$$E_{rr} = \frac{L^2}{2I_b} + L_a^2\left(\frac{1}{2I_a} - \frac{1}{2I_b}\right) \tag{7.27}$$

where L^2 is the square of the angular momentum. By replacing any classical quantity by its corresponding quantum expectation value, we immediately obtain the *quantum rotational energy E_{rot} of a symmetric top molecule*

$$E_{rot} = \frac{l(l+1)\hbar^2}{2I_b} + (m_l\hbar)^2\left(\frac{1}{2I_a} - \frac{1}{2I_b}\right) \tag{7.28}$$

where l is the angular quantum number and $m_l\hbar$ is the projection of **L** on the symmetry axis of the molecule, reflecting the typical quantization rule of any angular momenta. As usual, $m_l = -l, -l + 1,\dots, l - 1, l$ and, therefore, there is a twofold degeneracy in the rotational levels since E_{rot} depends on m_l^2.

7.3.2 Vibrations

Let us consider a polyatomic molecule containing N atoms. In total it has $3N$ nuclear degrees of freedom: three of them are associated with the motion of the molecular centre-of-mass, the other three describe the molecule as a rigid rotator, and the remaining $3N - 6$ ones are called the *vibrational degrees of freedom*[8].

Because of vibrations, each atom is displaced from its equilibrium position and, therefore, we can attempt to write *the classical nuclear potential energy of a polyatomic molecule* in terms of atomic displacements, by generalizing the procedure developed in section 7.1 To this aim we need to better define our notation. First of all, we recall that **R** indicates the full set of atomic positions and, accordingly, $\mathbf{R}_0$ will specify the equilibrium nuclear configuration of the molecule. Next, since atoms are labelled by Greek letters, we agree to indicate by $U_{\alpha,x} = R_{\alpha,x} - R_{\alpha,x}^{(0)}$ the x-component of the *displacement* of the αth atom (similar definitions hold for the y- and z-component of the same displacement vector $\mathbf{U}_\alpha$). This notation is really too pedantic for the purposes of our simplified discussion. Therefore, we agree to collapse the two indices identifying the αth atom and any Cartesian component of its displacement into just a single index: $(\alpha, x) \rightarrow i$ and similarly for (α, y) and (α, z).

---

[8] In linear molecules there are just two rotational degrees of freedom and, therefore, the resulting number of vibrations is $3N - 5$. This count is consistent with our treatment of the vibrations in a diatomic molecule (where $N = 2$) which was modelled as a one-dimensional oscillator.

This convention allows us to label the $3N$ degrees of freedom corresponding to the three Cartesian coordinates of the N displacements by collective Latin letters $i, j = 1, 2, 3, \dots, 3N$.

Now that the indexing notation has been clarified, we can proceed by applying the harmonic approximation to the classical nuclear potential energy $V_{nn}(\mathbf{R})$

$$V_{nn}(\mathbf{R}) \simeq V_{nn}(\mathbf{R}_0) + \sum_i \frac{\partial V_{nn}}{\partial U_i}\bigg|_{\mathbf{R}_0} U_i + \frac{1}{2}\sum_{i,j}\frac{\partial^2 V_{nn}}{\partial U_i \partial U_j}\bigg|_{\mathbf{R}_0} U_i U_j$$

$$= \frac{1}{2}\sum_{i,j} V_{nn,ij} U_i U_j \tag{7.29}$$

where the final expression is obtained by setting $V_{nn}(\mathbf{R}_0) = 0$ for convenience and by calculating $(\partial V_{nn}/\partial U_i)_{\mathbf{R}_0} = 0$ since the energy is minimum at equilibrium. The quantities

$$V_{nn,ij} = \frac{\partial^2 V_{nn}}{\partial U_i \partial U_j}\bigg|_{\mathbf{R}_0} \tag{7.30}$$

play the physical role of *effective force constants*. Those ones corresponding to the pure translational displacements and to the rigid rotations are of course zero. Therefore, *we understand that any sum over the i, j indices runs from 1 to $3N - 6$* (or to $3N - 5$ if we are considering a linear molecule). In other words, *hereafter the nuclear problem is purely vibrational*.

The tensorial character of the force constants appearing in equation (7.30) clearly indicates that atomic displacements mutually affect each other. The resulting dynamics is by far much more complex than in the diatomic molecule. By introducing the *mass weighted displacements*[9]

$$\bar{U}_i = M_i^{1/2} U_i \tag{7.31}$$

where M_i is the mass of the atom undergoing the displacement U_i, we can rewrite the total vibrational energy E_{vib} of the molecule as

$$E_{vib} = \frac{1}{2}\sum_i \left(\frac{d\bar{U}_i}{dt}\right)^2 + \frac{1}{2}\sum_{i,j}\frac{V_{nn,ij}}{(M_i M_j)^{1/2}} \bar{U}_i \bar{U}_j \tag{7.32}$$

This equation cannot be separated because of the $V_{nn,ij}$ terms with $i \neq j$. A standard procedure of classical mechanics [7] consists in replacing the weighted displacements by a new set of variables Q_i which are given in the form of the following linear combinations

---

[9] It should be remembered that i contains information either on the Cartesian coordinate of the displacement vector or on the specific atom of the molecule.

$$Q_i = \sum_l \Lambda_{il} \bar{U}_l \tag{7.33}$$

where the coefficients Λ_{il} are suitable constants related to the atomic masses of the chemical species forming the molecule and to the effective force constants governing the vibrations. This allows us to rewrite the vibrational energy in the more convenient form

$$E_{\text{vib}} = \frac{1}{2}\sum_i \left(\frac{dQ_i}{dt}\right)^2 + \frac{1}{2}\sum_{i,j} \lambda_i \, Q_i^2 \tag{7.34}$$

where the λ_i coefficients play once again the role of an (effective) force constant. The new coordinates Q_i are *collective coordinates* since they do not describe the motion of a single atom; rather, *they describe a displacement pattern collectively affecting all atoms in the molecule.* The corresponding oscillations are named *normal vibrational modes.* In figure 7.7 we show some of them in two different triatomic molecules.

The classical energy described by equation (7.34) can be straightforwardly quantized into a vibrational Hamiltonian operator

$$E_{\text{vib}} \rightarrow \hat{H}_{\text{vib}} = \sum_i \hat{h}_i \quad \text{with} \quad \hat{h}_i = -\frac{\hbar^2}{2}\frac{\partial^2}{\partial Q_i^2} + \frac{1}{2}\lambda_i Q_i^2 \tag{7.35}$$

describing *the complex of all molecular vibrations as the sum of* $3N - 6$ *independent one-dimensional harmonic oscillators.* By treating each of them as shown in appendix J, we easily obtain the quantum vibrational energy (in harmonic approximation) of a polyatomic (non-linear) molecule

$$E_{\text{vib}} = \sum_i \left(n_i + \frac{1}{2}\right) \hbar\omega_i \tag{7.36}$$

where n_i are the vibrational quantum numbers. The total vibrational wavefunction is the product of normalized functions given in equation (J.7). We finally observe that the above quantum oscillators are not referred to single atoms, but to normal vibrational modes.

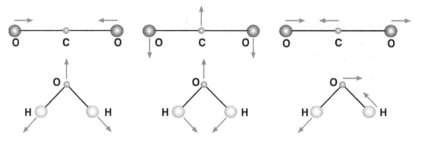

Figure 7.7. Some normal vibrational modes in CO_2 and H_2O molecules. Red arrows indicate the atomic displacements.

Further reading and references

[1] Bransden B H and Joachain C J 1983 *Physics of Atoms and Molecules* (Harlow: Addison-Wesley Longman)

[2] Demtröder W 2010 *Atoms, Molecules and Photons* (Heidelberg: Springer)

[3] Rigamonti A and Carretta P 2009 *Structure of Matter* 2nd edn (Milano: Springer)

[4] Atkins P and Friedman R 2011 *Molecular Quantum Mechanics* 5th edn (Oxford: Oxford University Press)

[5] Glazer M and Wark J 2001 *Statistical Mechanics: A Survival Guide* (Oxford: Oxford University Press)

[6] Colombo L 2022 *Statistical Physics of Condensed Matter Systems Physics: A Primer* (Bristol: IOP Publishing)

[7] Goldstein H 1996 *Classical Mechanics* 2nd edn (Reading, MA: Addison-Wesley)

IOP Publishing

Atomic and Molecular Physics (Second Edition)
A primer
Luciano Colombo

Chapter 8

Electronic structure of molecules

Syllabus—By means of the linear combination of atomic orbitals approach, we will develop the notion of 'molecular orbital', indeed the most fundamental concept in molecular physics. Then, we will discuss the electronic configurations of diatomic and polyatomic molecules, exploiting the fundamental role played by symmetry. A number of case studies will be explicitly addressed, highlighting the most important electronic features in molecules. As an example of a quantitative approach to calculate the molecular electronic structure, we outline the extended Hückel model which, although outdated by modern quantum chemistry methods, is still pedagogically relevant. Finally, within the conceptual framework assigned by the Franck–Condon principle, we will discuss the physics of molecular transitions when the full set of nuclear and electronic degrees of freedom is properly taken into account.

8.1 Problem definition

The theoretical framework for the determination of the molecular electronic structure was elaborated in section 6.2, fully exploiting the Born–Oppenheimer approximation. It is worth recalling the two key features of this conceptual scheme. First of all, it should be remembered that the electron energy eigenvalue problem, formulated by means of equation (6.6), must be solved separately for each nuclear configuration $\mathbf{R}$, i.e., electrons are described as adiabatically following the comparatively much slower nuclear dynamics; therefore, both their total energy $E_e(\mathbf{R})$ and total wavefunction $\psi_e^{(\mathbf{R})}(\mathbf{r})$ depend on $\mathbf{R}$ only parametrically[1]. Next, our arguments have been developed by completely neglecting magnetic and relativistic effects. In other words, only electrostatic electron–electron and electron–nucleus interactions have been taken into account.

---

[1] As in section 6.2 we will indicate by $\mathbf{r}$ and by $\mathbf{R}$ the full set of coordinates for electrons and nuclei, respectively.

While assuming to work below the approximations mentioned above, the problem defined in equation (6.6) is still formidably complicated, except for the simplest molecules (like, e.g., the hydrogen molecular ion H^+ or the elementary H_2 molecule). Basically, all of the modern techniques developed to solve equation (6.6) extensively make use of numerical calculations: this defines an independent advanced field of research and applications usually referred to as *quantum or computational chemistry* [1–4]. Computer simulations are typically performed at two different levels of sophistication, respectively corresponding to *ab initio* or semi-empirical methods. In the first case, a theoretical model is assumed (for instance, the Hartree–Fock or the configuration interaction one) and the corresponding quantum equation is self-consistently solved by numerical techniques without any other input than the atomic number of the nuclei present in the molecule. This approach has two main strengths: (i) it allows reaching the so-called *chemical accuracy*, which consists of predicting molecular energies within a few hundredths of eV with respect to the experimental data and (ii) it is transferable, i.e., it can be in principle applied to any molecule. The price to pay is a huge computational effort, making these calculations either in need of state-of-the-art computing resources and, in any case, very heavy, if not simply out-of-reach. This state of affairs justifies the use of semi-empirical methods: they adopt simplified forms of the Hamiltonian appearing in equation (6.6) and/or make use of adjustable parameters calibrated on experimental data (like, e.g., adjustable Coulomb, overlap, or resonance integrals as defined below). In this case, the computational workload is significantly reduced, but transferability and accuracy must be established for each specific application.

It is out of place to go into details of such a huge, very technical, and advanced topic. Adopting the same pedagogical approach as in the previous chapters, we will rather develop a *phenomenological approach* for the molecular electronic structure. More specifically, we will elaborate first the concept of *molecular orbital*, indeed the basic building block of any molecular theory. Next, we will develop some new basic features by studying diatomic molecules either by taking into account symmetry principles and by introducing the spin at the minimum theoretical level needed to fulfil the Pauli principle. The general features so elaborated will be eventually extended to polyatomic molecules, drawing a qualitative picture in some relevant case.

8.2 Molecular orbitals

The simplest molecule is the hydrogen molecular ion H_2^+, which will be here adopted as the case study to elaborate the key concept of *molecular orbital*. Its elementary structure is shown in figure 8.1, together with the frame of reference for the definition of any position or distance vector. In this case the quantum mechanical problem for the lone electron takes the form

$$\left(-\frac{\hbar^2}{2m_e}\nabla^2 - \frac{1}{4\pi\varepsilon_0}\frac{e^2}{r_A} - \frac{1}{4\pi\varepsilon_0}\frac{e^2}{r_B} \right)\psi(\mathbf{r}) = E\psi(\mathbf{r}) \tag{8.1}$$

where we used a simplified notation omitting the subscript 'e' from the wavefunction and the energy symbols, as well as the indication of the parametric dependence of

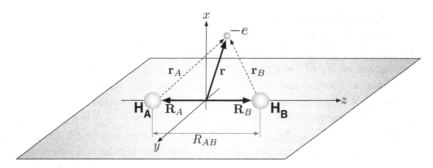

Figure 8.1. The hydrogen molecular ion H_2^+. All position and distance vectors used in the text are here indicated.

such quantities on the internuclear distance $R_{AB} = |\mathbf{R}_A - \mathbf{R}_B|$. This is the specific form of equation (6.6) for the H_2^+ molecule. Looking for its solution[2], we observe that if R_{AB} becomes very large then the molecular wavefunction should not significantly differ from the atomic ψ_{1s} one, indifferently referred to the nucleus A or to the nucleus B, depending on where the electron is actually located. This situation is named HH^+ *configuration*: it corresponds to a neutral hydrogen atom placed at infinite distance from a single proton. It is, therefore, very reasonable to write the trial solution of equation (8.1) as the *linear combination of two ψ_{1s} atomic orbitals* (we understand that we are interested in the molecular ground state)

$$\psi(\mathbf{r}) = c_1 \, \psi_{1s}^A(\mathbf{r}_A) + c_2 \, \psi_{1s}^B(\mathbf{r}_B) \tag{8.2}$$

where $c_{1,2}$ are constants to be determined, while $\psi_{1s}^A(\mathbf{r}_A)$ is a $1s$-like atomic orbital centred at the position of the nucleus A and calculated at the actual electron position $\mathbf{r}_A = \mathbf{R}_A - \mathbf{r}$ (see figure 8.1); a similar definition is obviously valid for $\psi_{1s}^B(\mathbf{r}_B)$. They will be hereafter referred to as *LCAO wavefunctions* (where the acronym LCAO stands for 'linear combination of atomic orbitals').

In order to calculate $c_{1,2}$ it is useful to follow a variational approach as outlined in section 5.4.1, which in the present case consists in two steps: first (i) we calculate the energy E as the expectation value of the Hamiltonian operator appearing in equation (8.1) on the trial solution given in equation (8.2), and then (ii) we minimize such an energy with respect to the c_1 and c_2 parameters, eventually obtaining the electronic molecular energy for the ground state. The tricky point is that the trial function has not been normalized yet. Therefore, the energy E is rigorously defined as

$$E = \frac{\int \psi(\mathbf{r})^* \hat{H} \psi(\mathbf{r}) \, d\mathbf{r}}{\int \psi(\mathbf{r})^* \psi(\mathbf{r}) \, d\mathbf{r}} \tag{8.3}$$

where we have set

---

[2] We remark that, by adopting suitable elliptic coordinates, this equation can be solved exactly, as shown in [5]. We nevertheless prefer to follow the variational approach for further convenience.

$$\hat{H} = -\frac{\hbar^2}{2m_e}\nabla^2 - \frac{1}{4\pi\varepsilon_0}\frac{e^2}{r_A} - \frac{1}{4\pi\varepsilon_0}\frac{e^2}{r_B} \qquad (8.4)$$

By inserting equation (8.2) into equation (8.3) we obtain after some straightforward algebra

$$E(c_1, c_2) = \frac{1}{c_1^2 + c_2^2 + 2c_1c_2S_{AB}}\left[(c_1^2 + c_2^2)H_{AA} + 2c_1c_2\,H_{AB}\right] \qquad (8.5)$$

where we have used the abbreviations

$$H_{AA} = \int [\psi_{1s}^A(\mathbf{r})]^*\hat{H}\psi_{1s}^A(\mathbf{r})\,d\mathbf{r} = \int [\psi_{1s}^B(\mathbf{r})]^*\hat{H}\psi_{1s}^B(\mathbf{r})\,d\mathbf{r} = H_{BB}$$

$$H_{AB} = \int [\psi_{1s}^A(\mathbf{r})]^*\hat{H}\psi_{1s}^B(\mathbf{r})\,d\mathbf{r} = \int [\psi_{1s}^B(\mathbf{r})]^*\hat{H}\psi_{1s}^A(\mathbf{r})\,d\mathbf{r} = H_{BA} \qquad (8.6)$$

$$S_{AB} = \int [\psi_{1s}^A(\mathbf{r})]^*\psi_{1s}^B(\mathbf{r})\,d\mathbf{r} = \int [\psi_{1s}^B(\mathbf{r})]^*\psi_{1s}^A(\mathbf{r})\,d\mathbf{r} = S_{BA}$$

The *Coulomb integral* H_{AA} represents the energy of the electron when the molecule is in the HH$^+$ configuration[3] (and, similarly, H_{BB} is the energy of the H$^+$H configuration), while the last term S_{AB} is known as the *overlap integral*: it estimates the superposition of the two atomic orbitals. The mixed integral H_{AB} is called *electron resonance integral* and it has no classical counterpart. Their calculation [6–8] is boring but not difficult and leads to the following expression for the Coulomb integral

$$H_{AA} = E_{1s}^{(H)} - \underbrace{\frac{e^2}{4\pi\varepsilon_0}\frac{1}{R_{AB}}\left[1 - \left(1 + \frac{R_{AB}}{a_0}\right)\exp(-2R_{AB}/a_0)\right]}_{=E_{AA}} = E_{1s}^{(H)} - E_{AA} \qquad (8.7)$$

for the overlap integral

$$S_{AB} = \left[1 + \frac{R_{AB}}{a_0} + \frac{1}{3}\left(\frac{R_{AB}}{a_0}\right)^2\right]\exp(-R_{AB}/a_0) \qquad (8.8)$$

and for the electron resonance integral[4]

$$H_{AB} = E_{1s}^{(H)}S_{AB} - \underbrace{\frac{e^2}{4\pi\varepsilon_0 a_0}\left(1 + \frac{R_{AB}}{a_0}\right)\exp(-R_{AB}/a_0)}_{=E_{AB}} = E_{1s}^{(H)}S_{AB} - E_{AB} \qquad (8.9)$$

where we used $E_{1s}^{(H)}$ for the ground-state energy of atomic hydrogen and the Bohr radius a_0. Coulomb and overlap integrals are a positive and rapidly decreasing function of R_{AB}. For instance, the overlap integral calculated at $R_{AB} = 4a_0$ is as small as ~0.2.

---

[3] In order to obtain the *total molecular energy*—still without rotations and vibrations—we should add the nuclear term $e^2/4\pi\varepsilon_0 R_{AB}$.

[4] Even in this case we should add a term $S_{AB}\,e^2/4\pi\varepsilon_0 R_{AB}$ in order to include the nuclear contribution.

We now proceed with the second step of the variational procedure by setting the conditions for E being minimum

$$\frac{\partial E(c_1, c_2)}{\partial c_1} = 0 = \frac{\partial E(c_1, c_2)}{\partial c_2} \tag{8.10}$$

which lead to the following system of algebraic equations

$$\begin{cases} (H_{AA} - E)\, c_1 + (H_{AB} - S_{AB}E)\, c_2 = 0 \\ (H_{AB} - S_{AB}E)\, c_1 + (H_{AA} - E)\, c_2 = 0 \end{cases} \tag{8.11}$$

whose solutions are obtained by solving the determinant equation

$$\begin{vmatrix} H_{AA} - E & H_{AB} - S_{AB}E \\ H_{AB} - S_{AB}E & H_{AA} - E \end{vmatrix} = 0 \tag{8.12}$$

providing *two different energies*

$$E_{\rm b} = \frac{H_{AA} + H_{AB}}{1 + S_{AB}}$$

$$E_{\rm ab} = \frac{H_{AA} - H_{AB}}{1 - S_{AB}} \tag{8.13}$$

where the meaning of the indices 'b' (standing for *bonding*) and 'ab' (standing for *anti-bonding'*) will be explained soon. By replacing these solutions into equation (8.11) it is easy to prove that it can only be $c_1 = \pm c_2$. Accordingly, the corresponding normalized wavefunctions

$$\psi_{\rm b}(\mathbf{r}) = \frac{\psi_{1s}^A(\mathbf{r}_A) + \psi_{1s}^B(\mathbf{r}_B)}{\sqrt{2 + 2S_{AB}}}$$

$$\psi_{\rm ab}(\mathbf{r}) = \frac{\psi_{1s}^A(\mathbf{r}_A) - \psi_{1s}^B(\mathbf{r}_B)}{\sqrt{2 - 2S_{AB}}} \tag{8.14}$$

are named *bonding and anti-bonding molecular orbitals*. We remark once again that they have been obtained as a *linear combination of atomic orbitals*.

We are now ready to write the *total molecular energy* when it is accommodated on a quantum state described by LCAO molecular orbitals. By duly inserting the nucleus–nucleus interaction contribution, we easily obtain

$$E_{\rm b}(R_{AB}) = E_{1s}^{\rm (H)} + \frac{e^2}{4\pi\varepsilon_0 R_{AB}} - \frac{E_{AA} + E_{AB}}{1 + S_{AB}}$$

$$E_{\rm ab}(R_{AB}) = E_{1s}^{\rm (H)} + \frac{e^2}{4\pi\varepsilon_0 R_{AB}} + \frac{E_{AB} - E_{AA}}{1 - S_{AB}} \tag{8.15}$$

where we have put in evidence the R_{AB}-dependence. In figure 8.2 (left) we show how such energies vary upon the internuclear distance, once equations (8.7), (8.8), and

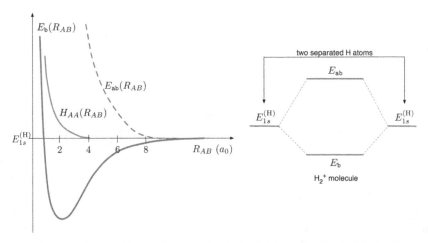

Figure 8.2. Left: the total energy of the hydrogen molecular ion H_2^+ in the bonding (b, full blue line) and anti-bonding (ab, dashed blue line) states as function of the internuclear distance R_{AB}; the energy H_{AA} of the molecule in the state HH^+ is shown for comparison (red line). Right: the corresponding energy diagram (not in scale).

(8.9) are used. The main emerging feature is that $E_{ab}(R_{AB})$ is always larger than the energy $E_{1s}^{(H)}$ of the HH^+ configuration: we conclude that *when the electron is accommodated in the anti-bonding level, the molecule cannot form a stable bound state*. This explains the use of the term 'anti-bonding'. In contrast, when the electron is described by the $\psi_b(\mathbf{r})$ wavefunction there exists a large interval of internuclear distances over which $E_b(R_{AB}) < E_{1s}^{(H)}$ and, therefore, the H_2^+ molecule forms: it is a truly 'bonding' situation. The $E_b(R_{AB})$ curve shows a minimum at $R_{AB}^0 = 1.32$ Å, corresponding to the a total energy as small as $E_b(R_{AB}^0) = E_{1s}^{(H)} - 1.76$ eV. In spite of the really simple guess adopted for the trial function[5] given in equation (8.2), these results agree reasonably well with the experimental data for the equilibrium distance $R_{AB}^0 = 1.06$ Å and the dissociation energy 2.65 eV. Another important conclusion is that *the molecular bonding is ensured by the resonance integral H_{AB} rather than by the Coulomb H–H$^+$ interaction*. While, as anticipated, this energy term is purely quantum mechanical, we can imagine that it could describe a classical situation where the electron is shared between the two nuclei, forming a sort of 'resonant state' with either E_b or E_{ab} energy. This picture is consistent with the qualitative discussion developed in section 6.3.2 about the covalent bonding: as a matter of fact, the odd and even wavefunctions plotted in figure 6.2 correspond, respectively, to $\psi_{ab}(\mathbf{r})$ and $\psi_b(\mathbf{r})$. A more accurate classification of molecular orbitals according to symmetry is presented in the next section.

Finally, the right panel of figure 8.2 shows the molecular energy diagram as calculated within the present LCAO approximation. The diagram is not in scale in

---

[5] A better choice is building LCAO molecular orbitals by using hydrogenic 1s wavefunctions with $Z = 2$, reflecting the fact that for very small values of the internuclear distance R_{AB} the electron feels a nuclear Coulomb field very similar to the He$^+$ case.

order to visually highlight that *the $\psi_{ab}(\mathbf{r})$ state is more anti-bonding that the $\psi_b(\mathbf{r})$ is bonding*. It is worth remarking that the sole $1s$ atomic wavefunction has generated two possible molecular orbitals.

8.3 Electronic configurations

At this point in our discussion it should already be clear that the calculation of the molecular electronic structure is a really tough problem. Even the very simple case of the H_2^+ molecular ion has led to non-trivial formal developments, with the introduction of subtle new concepts carrying unexpected features. From this point on, we will proceed in a semi-quantitative way by rooting our entire discussion in the LCAO approximation. Since we are going to describe multi-electron systems, the spin degree of freedom will be added to our analysis in order to fulfil the Pauli exclusion principle. No other spin-related features (like e.g., spin–orbit or spin–spin interactions) will be taken into account. Symmetry, a key new feature underlying molecular physics, will also be introduced in our arguments as a guideline to classify molecular states. Finally, as in the case of nuclear motions, the fundamental concepts will be elaborated considering diatomic molecules and, only afterwards, they will they be extended to the polyatomic case.

Once again, we remark that the following discussion, although based on quantum theory, will only provide semi-quantitative rules of thumb for a rational identi-fication of the ground state electronic configuration of a molecule. Quantitative predictions can only be elaborated by numerical quantum chemistry methods [1–4].

8.3.1 The role of symmetry

Molecules are extended objects and, therefore, they benefit from symmetry proper-ties or, equivalently, *their physics is left unchanged by applying suitable symmetry operations* like, e.g., a rotation around an axis or the reflection through a plane or the inversion with respect to a point in the space. Symmetry properties apply to the space part of the electron wavefunction, which will be the only part addressed in this section. The most accurate formal language for describing the molecular symmetry is that of group theory [8] which, however, falls beyond the scope of this manual. Here we limit ourselves to introducing the fundamental concepts in the case of diatomic molecules, for which it is possible to elaborate an elementary, but nevertheless meaningful and useful, picture.

Any diatomic molecule is characterized by an *axial symmetry*: the physics of the molecule is invariant under rotations about the molecular axis, as shown in figure 8.3. According to classical mechanics we can state that the angular momentum $\mathbf{L}$ of each electron in the molecule is not a constant: it actually undergoes precession around the molecular axis. This reflects in quantum theory by stating that *the only good quantum number is the magnetic one* or, equivalently, by stating that the L_z component may have values $L_z = m_l \hbar$, with $m_l = 0, \pm 1, \pm 2, \ldots$ where we have identified the molecular axis as the z-direction.

The physics of any diatomic molecule is also invariant under *reflections through all planes containing the molecular axis*: for instance, with respect to the frame of

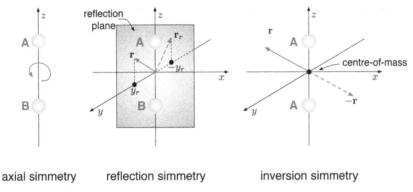

axial simmetry   reflection simmetry   inversion simmetry

Figure 8.3. The main symmetry properties of a diatomic molecule. The inversion symmetry holds only for homonuclear molecules with $A = B$.

reference shown in figure 8.3, the reflection through the xz plane corresponds to the coordinate change $y \to -y$. An electronic wavefunction corresponding to a state with $L_z = +m_l\hbar$ is converted upon reflection into a different one with $L_z = -m_l\hbar$. The two eigenstates, however, have the same energy (simply because the electron Hamiltonian commutes with the reflection operator[6]) and, therefore, *any molecular state with $m_l \neq 0$ is doubly degenerate.* In conclusion, the real good quantum number is $|m_l|$. By extending the labelling notation introduced in section 3.1.4 for electron shells in atoms, we will indicate the various molecular states $|m_l| = 0, \pm1, \pm2, \pm3,\ldots$ by the Greek letters σ, π, δ, ϕ,..., respectively. The only σ state is non-degenerate, while all the other ones have a twofold degeneracy.

We finally observe that homonuclear diatomic molecules benefit from an additional symmetry, namely: the *inversion symmetry through the molecular centre-of-mass* (which coincides with the origin of the frame of reference in figure 8.3 in the specific $A = B$ case). In formal language: the Hamiltonian operator is invariant upon the change $\mathbf{r} \to -\mathbf{r}$ of the electron coordinates. Its eigenfunctions can only be either *even* $\psi(\mathbf{r}) = \psi(-\mathbf{r})$ or *odd* $\psi(\mathbf{r}) = -\psi(-\mathbf{r})$ upon reflection, since it must always hold that $|\psi(\mathbf{r})|^2 = |\psi(-\mathbf{r})|^2$. In molecular physics it is customary to label them as *gerade (g)* or *ungerade (u)*, respectively, using the corresponding German words.

Molecular orbitals are built according to the LCAO principle and classified according to symmetry properties: the result is qualitatively shown in figure 8.4 (left). Conceptually, a two-step procedure is followed: (i) two infinitely-distant atoms are progressively approached; and (ii) during their approach atomic orbitals are differently combined, generating all possible molecular orbitals. We remark that the procedure illustrated in figure 8.4 (left) is valid for a homopolar molecule and that only the angular distribution of the wavefunction is shown, modulated by its radial part. Molecular orbitals are classified by the symbol σ, π, δ, ϕ,... followed by (nl) which indicates the principal and angular quantum numbers of the atomic orbitals used in the linear combination. The parity index is attached as a g or u label and, finally, the

---

[6] We can corroborate this formal argument with a classic parallel. The sign of m_l determines the sense of precession of $\mathbf{L}$ around the molecular axis, a feature that obviously leaves the electron energy unaffected.

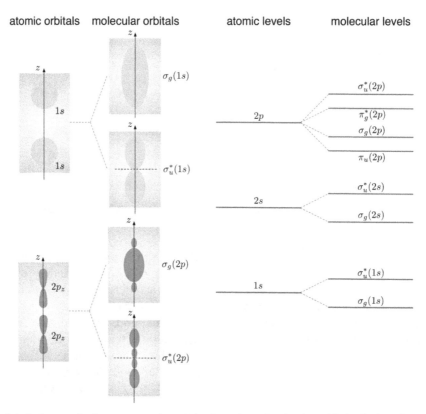

Figure 8.4. Left: a qualitative representation of the formation of molecular orbitals obtained from linear combinations of 1s and $2p_z$ atomic wavefunctions is shown upon projection onto a plane containing the molecular axis z. Right: the qualitative energy diagram (not in scale) of the resulting diatomic molecule (we remark that π molecular states are obtained through the combination of $2p_x$ or $2p_y$ atomic wavefunctions, not shown in the left panel).

indication of the anti-bonding character is reported by an asterisk '*' symbol. Interestingly enough, anti-bonding orbitals display a nodal plane normal to the molecular axis (dashed line in figure 8.4). It is useful to consider the case of the H_2^+ molecule to see how this complicated notation works: the bonding $\psi_b(\mathbf{r})$ molecular orbital is labelled $\sigma_g(1s)$, while the anti-bonding $\psi_b(\mathbf{r})$ is labelled $\sigma_u^*(1s)$. Figure 8.4 (left), although qualitative, provides evidence of some general trend. First, we observe that *atomic states of the same energy give rise to the lowest bonding molecular orbitals* and, therefore, in very good approximation we can neglect the mixed combinations (like, for instance, the superposition of 1s–2s, 1s–2p, … states). Next, we remark that *ungerade* states can be the lowest-energy level within a single set, contrary to an expectation possibly misguided by purely aesthetic arguments, as shown in the right panel of figure 8.4 where the corresponding qualitative energy diagram is reported.

8.3.2 Diatomic molecules

8.3.2.1 The H_2 molecule

Let us start with the simplest homonuclear molecule, namely H_2, which we will use as a study guideline to elaborate the main concepts. Within a Hartree-like approach, the space part of the wavefunction describing its ground state is obtained by simply adding a second $\sigma_g(1s)$ electron to the H_2^+ minimum energy configuration: if $\mathbf{r}_1$ and $\mathbf{r}_2$ indicate the positions of the two electrons, we can write in compact notation such a total wavefunction as $\psi_{\sigma_g(1s)}(\mathbf{r}_1)\psi_{\sigma_g(1s)}(\mathbf{r}_2)$. This function is symmetric under electron interchange and, therefore, in order to fulfil the Pauli principle we must multiply it by an *antisymmetric* spin function χ_A of the form given in equation (5.14). This leads to the *full form of the wavefunction describing the H_2 ground state*

$$\psi_{GS}^{(H_2)}(\mathbf{r}_1, \mathbf{r}_2) = \frac{1}{\sqrt{2}} \, \psi_{\sigma_g(1s)}(\mathbf{r}_1)\psi_{\sigma_g(1s)}(\mathbf{r}_2) \left[\chi_\uparrow(1)\chi_\downarrow(2) - \chi_\downarrow(1)\chi_\uparrow(2) \right] \tag{8.16}$$

The corresponding energy $E_{GS}^{(H_2)}$ is calculated as the expectation value -on such a state- of the total Hamiltonian

$$\hat{H} = -\frac{\hbar}{2m_e} \sum_{i=1,2} \nabla_i^2 + \frac{e^2}{4\pi\varepsilon_0}\left[-\sum_{i=1,2}\sum_{\alpha=A,B} \frac{1}{|\mathbf{r}_i - \mathbf{R}_\alpha|} + \frac{1}{|\mathbf{r}_1 - \mathbf{r}_2|} + \frac{1}{|\mathbf{R}_A - \mathbf{R}_B|} \right] \tag{8.17}$$

where $\mathbf{r}_{1,2}$ and $\mathbf{R}_{A,B}$ are, respectively, the electron and nuclei positions with respect to the origin of frame of reference placed at the molecule centre-of-mass. This laborious calculation [6, 7] leads to an equilibrium internuclear distance $R_{AB}^0 = 0.84$ Å at which we have $E_{GS}^{(H_2)} = 2E_{1s}^{(H)} - 2.68\text{eV}$. The molecular dissociation energy is accordingly predicted as large as $E_{diss} = 2.68$ eV. These results are in fair agreement with the experimental data, respectively $R_{AB}^{0,\text{expt}} = 0.74$ Å and $E_{diss}^{\text{expt}} = 4.75$ eV. There is room for improvement, which is natural to look for in the direction of a better description of the space part of the wavefunction $\psi_{GS}^{(H_2)}(\mathbf{r}_1, \mathbf{r}_2)$: instead of guessing a Hartree-like wavefunction, we could use *a linear combination of molecular orbitals products*, including mixed blends like, e.g., $\psi_{\sigma_g(1s)}(\mathbf{r}_1)\psi_{\sigma_g(2s)}(\mathbf{r}_2)$ or $\psi_{\sigma_g(2s)}(\mathbf{r}_1)\psi_{\sigma_g(2s)}(\mathbf{r}_2)$ or other ones, in addition to the $\psi_{\sigma_g(1s)}(\mathbf{r}_1)\psi_{\sigma_g(1s)}(\mathbf{r}_2)$ term used before. This is the essence of the *molecular configuration interaction method* [8].

8.3.2.2 Homonuclear molecules

The ground state configuration of other homonuclear diatomic molecules can be identified by applying the basic principles so far developed: (i) atomic orbitals forming a molecular one must be shaped so as to ensure substantial overlap; (ii) the combination of atomic orbitals with similar energy must be preferred; and (iii) a maximum of two paired electrons can be accommodated on each molecular orbital, so as to fulfil Pauli exclusion (and, in case there are fewer electrons than can be accommodated on the orbital, they organize in such a way as to maximize the total spin). This qualitative approach allows us to define an additional general feature: *the* *stability of a homonuclear diatomic molecule depends on the relative number of electrons in bonding and anti-bonding states.* More specifically, a stable configuration is found provided the electrons in bonding orbitals exceed in number the electrons in anti-bonding ones. In addition, *stability decreases as the difference between the two electron populations decreases.* For instance, the ground state of He_2 has two paired electrons on both the $\sigma_g(1s)$ and the $\sigma_u^*(1s)$ orbitals: the molecule is unstable and helium only exists as an atomic gas. The same conclusion is drawn for any other diatomic molecule formed by noble gas atoms. For the same reason, the molecular stability decreases, as shown in figure 8.5, along the series $N_2 \rightarrow O_2 \rightarrow F_2$ since the difference in the number of electrons in bonding and anti-bonding states consistently decreases. We remark that σ-states can accommodate just two paired electrons since $m_l = 0$. On the other hand, π-states can accommodate up to four electrons: they are doubly degenerate since their energy does not depend on the sign of the m_l quantum number (the same is valid as well for δ-, ϕ-, ... states). This allows in B_2 and O_2 to have two unpaired electrons in the highest-energy incomplete molecular orbital.

In figure 8.5 the *molecular spectroscopic notation* has been adopted to label the molecular ground state, defined as follows. The total electronic angular momentum $\mathbf{L}_{tot}$ is projected along the molecular axis to obtain its component $L_{tot,z} = m_{l_{tot}}\hbar$ with $m_{l_{tot}} = \sum_i m_{l_i}$ (of course the sum runs over the electrons, labelled by the i index). The

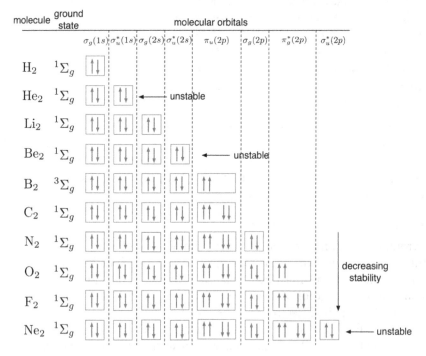

Figure 8.5. Ground state electronic configuration of some homopolar diatomic molecules. Each box represents a molecular state that can be filled by electrons with either spin up (↑) or spin down (↓).

energy depends on the absolute values $|m_{l_{tot}}| = 0, 1, 2, 3,...$ which are conventionally indicated as Σ, Π, Δ, Φ,.... The spin multiplicity $2S_{tot} + 1$ is added as a left-top superscript, while the g or u parity index is displayed as a right-bottom subscript.

8.3.2.3 *Heteronuclear molecules*

Following the discussion reported in section 6.3, we will assume that electrons in closed atomic shells are basically unaffected by the presence of the second atom and, therefore, they do not participate in the chemical bond (frozen core approximation). We will further simplify the problem by also assuming that *paired electrons in incomplete shells are also chemically inactive*. These assumptions greatly reduce the complexity of the molecular problem, as clearly evident in the NaCl case study: just the $3s$ electron of the Na atom and one $3p$ electron of the Cl atom will be considered in forming a molecular orbital.

Since the two nuclei are different, we expect that a heteronuclear diatomic molecule is characterized by a *non-uniformly distributed electron charge*. Accordingly, the trial LCAO function takes the most general form

$$\psi(\mathbf{r}) = c_1 \psi^A(\mathbf{r}_A) + c_2 \psi^B(\mathbf{r}_B) \tag{8.18}$$

where the two coefficients $c_{1,2}$ can no longer be equal (in absolute value) as in the homonuclear case. More specifically, the larger the electronegativity of an atom, the larger is its weight-factor c_i ($i = 1, 2$) in the linear combination.

Another general feature is that *in heteronuclear molecules the molecular orbitals are less shifted in energy with respect to atomic ones* than for homonuclear species. The argument to prove this statement is simple. By extending the formal development that has led to equation (8.12), we obtain a determinant equation of the form

$$\begin{vmatrix} H_{AA} - E & H_{AB} - S_{AB}E \\ H_{AB} - S_{AB}E & H_{BB} - E \end{vmatrix} = 0 \tag{8.19}$$

which translates into a second-order equation for the energy

$$(1 - S_{AB})^2 E^2 + (2S_{AB}H_{AB} - H_{AA} - H_{BB})E + (H_{AA}H_{BB} - H_{AB}^2) = 0 \tag{8.20}$$

where in this case $H_{AA} \neq H_{BB}$. If we assume that the overlap S_{AB} and the resonance integrals are negligibly small (i.e., $S_{AB} \sim 0$ and $H_{AA} - H_{BB} \gg H_{AB}$) as a consequence of the different chemical nature of the two atoms forming the molecule, we easily obtain the general expression of the energy of the bonding and anti-bonding states as

$$E_b = H_{AA} - \frac{H_{AB}^2}{H_{BB} - H_{AA}}$$
$$E_{ab} = H_{BB} + \frac{H_{AB}^2}{H_{AA} - H_{BB}} \tag{8.21}$$

where we have assumed $H_{BB} > H_{AA}$. Under the above postulations, we obtain $E_b \sim H_{AA}$ and $E_{ab} \sim H_{BB}$ as anticipated.

The general procedure outlined in the previous section can in principle be applied to heteronuclear molecules as well. It is a widely used notation to indicate the *highest occupied molecular orbital* by the acronym HOMO and the *lowest unoccupied molecular orbital* as LUMO. The energy difference between these levels is referred to as the *HOMO–LUMO gap*. Some molecular optical properties are extensively affected by its value. In molecular solids it plays a similar role to the gap between the valence and the conduction bands in crystalline semiconductors.

8.3.3 Polyatomic molecules

Based on the notions developed so far, it is really difficult to elaborate a quantitative description of the molecular electronic structure beyond the diatomic case: for any polyatomic species a computer-based approach is really needed. Let us for example consider the development of the LCAO method in the case of a polyatomic molecule in which $N \gg 2$ electrons participate in the formation of chemical bonds. The secular problem given by equation (8.21) results in the diagonalization of an $N \times N$ matrix, a task that can be efficiently accomplished only by numerical methods. Furthermore, the number of Coulomb, overlap, and resonance integrals to be computed soon becomes overwhelmingly large, as N increases. In implementing the numerics needed to solve the secular problem, a number of advanced theoretical tools are used, among which the *group theory* plays a major role. It allows due consideration to be given to the symmetry of the molecule and, for instance, makes it possible to use symmetry-adapted coordinates rather, than Cartesian ones [8]. As outlined at the beginning of this chapter, numerical calculations can be implemented with either *ab initio* or semi-empirical methods.

In the attempt to provide at least some qualitative information about how to guess the main bonding characteristics in a polyatomic molecule, we recognise the main role played by the twofold *maximum orbital overlap principle*: (i) a bond between two atoms in a polyatomic molecule occurs along the direction of maximum overlap between the participating atomic wavefunctions; (ii) the larger the overlap, the greater the bond strength. Both statements are recognisably based on the arguments developed in the diatomic case. The quantum mechanical foundation of this empirical principle is offered by the so-called *quantum theory of atoms in molecules*, which provides the most formal, robust, predictive, and computable definition of 'chemical bond' [9].

By applying this principle we can draw a qualitative picture about the *electronic distribution* of a molecule. Let us consider the case of water, indeed a molecule of paramount importance in physics, earth sciences, (bio)chemistry, and astronomy. H_2O consists of three nuclei and ten electrons. As shown in figure 5.6 just two electrons of the oxygen atom are unpaired and placed in an incomplete sub-shell: in the line of our discussion above, we will consider only them in the process of formation of the H–O bond and we will describe them by means of a $2p_x$ and a $2p_y$ orbital, respectively. According to the maximum orbital overlap principle, we expect that the two H atoms should be placed on the x and y axis at the same distance from the O atom, as shown in figure 8.6 (top). Qualitatively, this prediction is confirmed

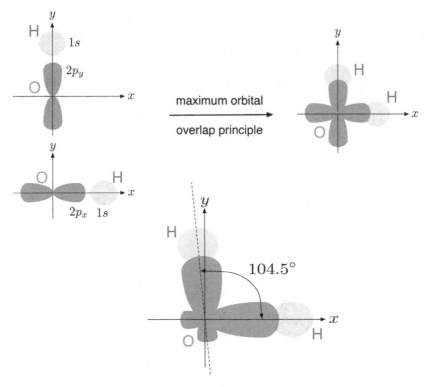

Figure 8.6. Top: the qualitative formation process of the bonds in the H_2O molecule, according to the maximum orbital overlap principle. Bottom: the actual electronic distribution of the molecule (in a magnified scale for greater clarity). The electronic distribution is projected on the xy plane.

by the experimental evidence: H_2O is not a linear molecule. However, its true structure differs from our qualitative prediction in two respects: (i) the bond angle[7] is 104.5° (or, equivalently, the two bonds are not normal); and (ii) the p-like oxygen orbitals are deformed with respect to their atomic counterparts. Both features—which can be accounted for only by a direct numerical quantum chemistry calculation—are due to the presence of the two nearby hydrogen atoms: they repel each other, thus widening the bond angle, and polarize the oxygen p-orbitals which assume the shape shown in figure 8.6 (bottom) where the true electronic distribution in the water molecule is illustrated. The H_2O polarization is confirmed by the experimental measurement of an electric dipole moment as large as 1.87 debye.

The application of the maximum orbital overlap principle can be extended to the case of more complex molecular architectures, although with increasing difficulty.

8.3.4 Orbital hybridization

It is important to address the case of carbon-based molecules, which play a key role either in basic science or in applied (bio)chemistry.

---

[7] The 'bond angle' is the angle formed by the directions of the two O–H bonds.

Carbon is a special case because of its very peculiar ground state electronic configuration $(1s)^2(2s)^2(2p)^2$ (see figure 5.6). Since the energy difference between the $2s$ and $2p$ states is small, at room temperature the excited configuration $(1s)^2(2s)^1(2p)^3$ is most likely found. In this case there are *four unpaired electrons in the highest $n = 2$ partially-filled shell*. There are, therefore, four atomic orbitals to play with in building molecular orbitals. Interesting enough, the three p-like atomic orbitals are directional, while the remaining s-like one is not: this makes it difficult to identify the bonding directions within the molecules. Furthermore, carbon is found to form up to four bonds.

The observed *tetravalent and directional bonding character* of carbon atoms is easily explained by assuming that *the four valence atomic orbitals are hybridized*. More formally: instead of using true $2s$ and $2p$ atomic orbitals in building the molecular ones, we rather make use of the following sp^3 *hybridized orbitals*[8]

$$\psi_1 = \frac{1}{2}\left(\psi_{2s} + \psi_{2p_x} + \psi_{2p_y} + \psi_{2p_z}\right)$$

$$\psi_2 = \frac{1}{2}\left(\psi_{2s} + \psi_{2p_x} - \psi_{2p_y} - \psi_{2p_z}\right)$$

$$\psi_3 = \frac{1}{2}\left(\psi_{2s} - \psi_{2p_x} + \psi_{2p_y} - \psi_{2p_z}\right) \tag{8.22}$$

$$\psi_4 = \frac{1}{2}\left(\psi_{2s} - \psi_{2p_x} - \psi_{2p_y} + \psi_{2p_z}\right)$$

These hybridized orbitals do not describe states with a well-defined value of angular momentum, but they are directional: their maxima point to the four vertexes of a tetrahedron, forming angles of about 109°. If we now consider the CH_4 (methane) molecule, we can apply the maximum orbital overlap principle to the four sp^3 hybridized carbon orbitals and to the $1s$ orbital of each H atom, obtaining a prediction pretty well confirmed by experimental evidence: CH_4 is a tetrahedral molecule with a C atom at its centre (see figure 8.7). The same sp^3 hybridized orbitals are used to guess the structure of more complex hydrocarbons[9], like the C_2H_6 molecule (ethane, see figure 8.7): in this case the carbon atoms form a bond by overlapping two sp^3 orbitals aligned along the internuclear axis (a kind of bond which is referred to as σ-bond), while the six hydrogen atoms saturate the remaining tetrahedral bonds[10].

Carbon is able to give rise to hybridizations other than the sp^3 one: this property is called *carbon allotropy*. First, we define the sp^2 hybridization scheme

---

[8] The linear combination leading to the definition of the sp^3 hybrid orbitals is developed in analogy to what was done in section 3.1.5. This 'mixture' of atomic orbitals quantifies their distortion caused by the interactions among the atoms forming the molecule.

[9] Hydrocarbons are molecules formed just by C and H atoms.

[10] We remark that the concept of sp^3 hybridization applies as well to all atoms with the same electron configuration of the highest-energy partially-filled electronic shell, like e.g., Si and Ge.

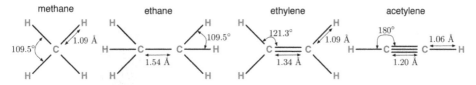

Figure 8.7. The graphical representation of some simple hydrocarbon molecules, with indication of single/double/triple bonds, bond lengths, and bond angles. Configurations have been obtained by adopting the more suitable hybridization scheme and the corresponding graph is projected onto the page plane.

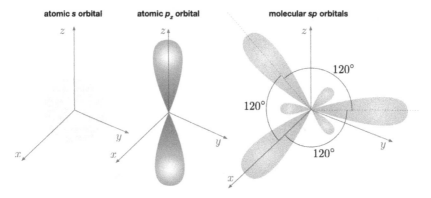

Figure 8.8. Shape of the molecular orbitals resulting from sp^2 hybridization (not in scale).

$$\psi_1 = \frac{1}{\sqrt{3}}\left(\psi_{2s} + \sqrt{2}\,\psi_{2p_x}\right)$$

$$\psi_2 = \frac{1}{\sqrt{3}}\left(\psi_{2s} - \frac{\psi_{2p_x}}{\sqrt{2}} + \frac{\sqrt{3}\,\psi_{2p_y}}{\sqrt{2}}\right) \qquad (8.23)$$

$$\psi_3 = \frac{1}{\sqrt{3}}\left(\psi_{2s} - \frac{\psi_{2p_x}}{\sqrt{2}} - \frac{\sqrt{3}\,\psi_{2p_y}}{\sqrt{2}}\right)$$

where the three orbitals lie on the xy plane and their maxima point to three directions forming angles of 120°, as shown in figure 8.8.

This hybridization scheme underlies the formation of planar hydrocarbons, like the C_2H_4 molecule (ethylene, see figure 8.7). In this case, the C–C bond is double, since it is formed by the overlap between a pair of sp^2 hybridized orbitals and a pair of $2p_z$ atomic orbitals (in both cases: one orbital for each C atom). The first overlap gives rise to a σ-bond, while the overlap between the $2p_z$ orbitals is said to form a π-bond. The graphical rendering of the two overlaps is reported in figure 8.9, while in the chemical representation the molecular single and double bonds are, respectively, shown as a single or double line, as found in figure 8.7.

σ bond   π bond

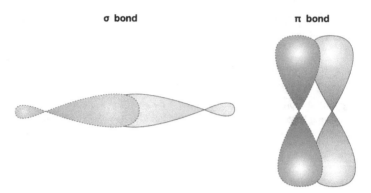

Figure 8.9. Left: overlap between sp^2 hybridized orbitals forming a σ-bond. Right: overlap between $2p_z$ atomic orbitals forming a π-bond.

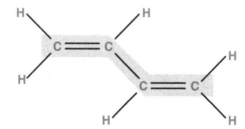

Figure 8.10. The butadiene molecule. The shaded area pictorially represents the molecular backbone.

Finally, we define the sp^1 hybridization scheme

$$\psi_1 = \frac{1}{\sqrt{2}}\left(\psi_{2s} + \psi_{2p_z}\right) \quad \psi_2 = \frac{1}{\sqrt{2}}\left(\psi_{2s} - \psi_{2p_z}\right) \tag{8.24}$$

which is useful to describe hydrocarbons with triple bonds, like the C_2H_2 molecule (acetylene, see figure 8.7). In this case the C–C bond is formed by the overlap of a pair of sp^1 hybridized orbitals (one for each carbon atom), giving rise to a σ-bond, and by the overlap of two $2p_x$ and two $2p_y$ atomic orbitals for each carbon atom, giving rise to two π-bonds. Triple bonds are graphically shown by triple lines.

The double possibility to generate σ- and π-bonds provides carbon with a unique versatility in forming molecules. An especially important class of C-based polyatomic species is represented by the *conjugated molecules*, which are planar hydrocarbons with three main characteristics: (i) they have a chain of σ-bonds formed by sp^2 hybridized orbitals, usually referred to as the *molecular backbone*; (ii) hydrogens are bonded to C atoms through the overlap of *s*-like and sp^2-like orbitals; and (iii) along the molecular backbone C atoms are further bonded by π-bonds which lie in the direction normal to the molecular plane. In figure 8.10 is shown the butadiene molecule as an example of this complex bonding environment. The key feature resulting from the above characteristics is that electrons accommodated on π-bonds, usually named 'π-electrons' (one for each C atom), are *delocalized along the molecular backbone* or, equivalently, they are not confined in the neighbourhood of any specific carbon atom. Under the action of an external electric field they are relatively free to move, which gives the conjugated molecule a very high electrical polarizability along the molecular backbone.

Conjugated molecules also exist as linear chains, as found in the case of *cumulene*, consisting of a sequence of double C=C bonds, and in the case of *polyyne*, consisting of an alternation of single C–C and triple C $\equiv$ C bonds. The different bonding structure reflects in different electric properties: cumulene is metallic, while polyyne is a semiconductor.

Finally, a special class of conjugated species is represented by *cyclic molecules* in which carbon atoms are placed at the vertexes of regular polyhedra. The result is a fascinating molecular architecture: the molecular backbone follows the polyhedron perimeter as a sequence of σ-bonds, hydrogens atoms are attached to each carbon atom through the overlap of *s*-like and sp^2-like orbitals and, finally, the remaining bonds are formed by π-electrons. In the specific case of the C_6H_6 molecule (benzene), the backbone is a regular hexagon along which six π-electrons are delocalized: they give rise to a sort of internal electric current which, in turn, is responsible for the strong diamagnetic behaviour of this molecule (a feature shared by all conjugated cyclic molecules). A more quantitative description will be offered in the next session.

In conclusion, carbon is a really very special atom whose chemistry is characterized by the fact that it is largely prone to hybridization. For instance, thanks to its allotropic character (and even without the assistance of H atoms), it gives rise to very diverse aggregates including diamond (an sp^3 hybrid) as well as graphene, nanotube structures, or fullerene molecules (all sp^2 hybrids) which are 3-, 2-, 1-, and 0-dimensional objects, respectively. Organic chemistry is, in turn, ruled over by carbon flexibility to form either σ- and π-bonds, which differently behave as for their ability to localize electrons. The combination of hybridization with the double character of the C–C bond makes biology possible (together with the presence of water). All these fascinating topics are extensively treated in organic chemistry [10] and bio-chemistry [11] textbooks.

8.3.5 A more quantitative approach: the extended Hückel method

In order to solve the electronic structure of some conjugated molecules in which single and double bonds alternate, a set of approximations can be introduced to make the problem exactly solvable[11]. More specifically, we address the mostly representative case of benzene, which can be thought of in two *resonant limit forms* shown in figure 8.11, suggesting that the real molecular geometry is actually intermediate (more specifically, as we shall see, it is a linear combination of the two forms). The Hückel method basically consists in treating separately the σ-bonds (which generate the rigid hexagonal structure of benzene) and the π-orbitals (where,

---

[11] This approach was introduced by E Hückel in 1931 and still represents the kernel of the so-called semi-empirical electronic structure methods [4]. Its counterpart in solid state physics is referred to as the tight-binding method [12].

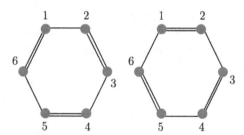

Figure 8.11. The two resonance limit forms of the benzene molecule. Blue dots are carbon atoms.

due to the symmetry of the problem, the role of each carbon atom is just the same). This assumption greatly simplifies the problems, since for each (i, j) pair of adjacent atoms (with $i, j = 1, 2,\dots , 6$), one can define an overlap integral S_{ij} and a resonance integral H_{ij} that do not depend on the specific position. It can also be assumed that the six Coulomb integrals H_{ii} are all equal.

Let us indicate by Ψ the total wavefunction describing the six electrons forming the π-bonds and by ϕ_i (with $i = 1, 2,\dots , 6$) the corresponding non-hybridized and normalised atomic p_z-orbitals (referring to figure 8.11, the z-axis is normal to the plane containing the hexagonal carbon ring). Then, we can write

$$\Psi = \sum_{i=1}^{6} c_i\phi_i \tag{8.25}$$

where the c_i's coefficients can be determined by minimizing the energy

$$\langle E\rangle_\Psi = \int \Psi^*\hat{H}\ \Psi dV \tag{8.26}$$

under the normalization constraint $\int \Psi^*\Psi dV = 1$. We remark that in the compact notation here used, the $\int ...dV$ symbol indicates the integral over all the electronic space coordinates and $\hat{H}$ represents the Hückel Hamiltonian operator for the π-electrons. By inserting equation (8.25) into equation (8.26) we get

$$\langle E\rangle_\Psi = \sum_{i}\sum_{j}H_{ij}c_i^*c_j \tag{8.27}$$

where $H_{ij} = \int \phi_i^*\hat{H}\phi_j dV$, while

$$\sum_{i}|c_i|^2 + \sum_{i}\sum_{j\neq i}S_{ij}c_i^*c_j = 1 \tag{8.28}$$

corresponds to the normalization constraint and $S_{ij} = \int \phi_i^*\phi_j dV$. To keep the formalism as simple as possible, we further assume that the c_i's coefficients are real and $H_{ji} = H_{ij}$.

The minimization condition and the normalization constraint are simultaneously solved by the method of the Lagrange multipliers which, after a little algebra, leads to

$$\sum_j (H_{ij} - \varepsilon S_{ij}) c_j = 0 \tag{8.29}$$

or equivalently

$$\begin{vmatrix} H_{11} - \varepsilon & H_{12} - \varepsilon S_{12} & H_{13} - \varepsilon S_{13} & H_{14} - \varepsilon S_{14} & H_{15} - \varepsilon S_{15} & H_{16} - \varepsilon S_{16} \\ H_{21} - \varepsilon S_{21} & H_{22} - \varepsilon & H_{23} - \varepsilon S_{23} & H_{24} - \varepsilon S_{24} & H_{25} - \varepsilon S_{25} & H_{26} - \varepsilon S_{26} \\ H_{31} - \varepsilon S_{31} & H_{32} - \varepsilon S_{32} & H_{33} - \varepsilon & H_{34} - \varepsilon S_{34} & H_{35} - \varepsilon S_{35} & H_{36} - \varepsilon S_{36} \\ H_{41} - \varepsilon S_{41} & H_{42} - \varepsilon S_{42} & H_{43} - \varepsilon S_{43} & H_{44} - \varepsilon & H_{45} - \varepsilon S_{45} & H_{46} - \varepsilon S_{46} \\ H_{51} - \varepsilon S_{51} & H_{52} - \varepsilon S_{52} & H_{53} - \varepsilon S_{53} & H_{54} - \varepsilon S_{54} & H_{55} - \varepsilon & H_{56} - \varepsilon S_{56} \\ H_{61} - \varepsilon S_{61} & H_{62} - \varepsilon S_{62} & H_{63} - \varepsilon S_{63} & H_{64} - \varepsilon S_{64} & H_{65} - \varepsilon S_{15} & H_{66} - \varepsilon \end{vmatrix} \begin{vmatrix} c_1 \\ c_2 \\ c_3 \\ c_4 \\ c_5 \\ c_6 \end{vmatrix} = 0 \tag{8.30}$$

where we made explicit the matrix form of these equations. By physical intuition, we further introduce a number of *Hückel approximations*: (i) the resonance integral H_{ij} is set to zero if atoms i and j are not first next neighbours[12]; (ii) for any (i, j) pair we set $H_{ij} = \beta$; (iii) all Coulomb integrals are set $H_{ii} = \alpha$; (iv) finally, it is guessed that $S_{ij} = 0$ by assuming that atomic orbitals centred on different atomic sites are orthogonal[13]. Under these approximations, equation (8.30) reduces to the much simpler form

$$\begin{vmatrix} \alpha - \varepsilon & \beta & 0 & 0 & 0 & \beta \\ \beta & \alpha - \varepsilon & \beta & 0 & 0 & 0 \\ 0 & \beta & \alpha - \varepsilon & \beta & 0 & 0 \\ 0 & 0 & \beta & \alpha - \varepsilon & \beta & 0 \\ 0 & 0 & 0 & \beta & \alpha - \varepsilon & \beta \\ \beta & 0 & 0 & 0 & \beta & \alpha - \varepsilon \end{vmatrix} \begin{vmatrix} c_1 \\ c_2 \\ c_3 \\ c_4 \\ c_5 \\ c_6 \end{vmatrix} = 0 \tag{8.31}$$

whose non-trivial solutions are obtained by imposing the determinant of the matrix to be null. This straightforwardly leads to six electron energy values

$$\varepsilon_1 = \alpha - 2\beta \qquad \varepsilon_{2,3} = \alpha - \beta \qquad \varepsilon_{4,5} = \alpha + \beta \qquad \varepsilon_6 = \alpha + 2\beta \tag{8.32}$$

which correspond to four different states, two of which are doubly degenerate. By placing the six π-electrons on them we predict

$$E_{\text{tot}}^{(H)} = 2(\alpha + 2\beta) + 4(\alpha + \beta) = 6\alpha + 8\beta \tag{8.33}$$

as *the total Hückel molecular energy* provided by the π-bonds only[14]. Let us now consider a resonant limit form (RLF) shown in figure 8.11 (it is irrelevant which

---

[12] This assumption (referred to as the first next-neighbours approximation) is justified by considering that the overlap between two orbitals centred on different atomic sites decays rapidly as the interatomic distance increases, due to the localized character of the p_z-orbitals.

[13] This condition (referred to as the orthogonality of the atomic basis set) can be in fact released: this makes the formalism a bit more complex, but it does not modify the reasoning we are going to develop. On the other hand, using a non-orthogonal basis set generally reflects in a higher accuracy; in this case the formalism is referred to as the extended Hückel method.

[14] The σ-bonds are not considered in this formulation of the Hückel theory: their role is simply to provide a stable hexagonal ring.

one): for each isolated double bond formed by the (i, j) atom pair, we should write the total wavefunction Ψ as

$$\Psi = c_i\phi_i + c_j\phi_j \tag{8.34}$$

understating that it describes the quantum state of just the two electrons coupled in that specific double bond. The corresponding bond energy is

$$E = \alpha \pm \beta \tag{8.35}$$

as shown in equation (8.13). By extending this argument to the other two double bonds present in the RLF molecule, we estimate a total molecular energy

$$E_{\text{tot}}^{(\text{RLF})} = 6(\alpha + \beta) \tag{8.36}$$

which is, however, larger than $E_{\text{tot}}^{(\text{H})}$ obtained by the Hückel theory. More precisely, we have

$$E_{\text{tot}}^{(\text{H})} - E^{(\text{RLF})} = 2\beta < 0 \tag{8.37}$$

which represents *the energy gain obtained by delocalizing the six π-electrons along the whole hexagonal backbone of the benzene molecule.*

8.4 Electronic transitions: the Franck–Condon principle

In the spirit of the Born–Oppenheimer approximation, we have so far separately discussed the nuclear motions and the electronic structure in molecules. It is now time to recover a unified vision, addressed to understanding transitions between different electronic levels. To keep the discussion at an elementary level, we will focus on the case of diatomic molecules.

Our knowledge indicates that, whenever a diatomic molecule undergoes a transition from the ground state electronic configuration to an excited state one, its total energy profile $E_{\text{T}}(\mathbf{R}) = E_e^{(\mathbf{R})} + E_n(\mathbf{R})$ changes and the internuclear distance $R = |\mathbf{R}|$ increases. More rigorously, we can write

$$E_{\text{T}} = E_e + E_{\text{rot}} + E_{\text{vib}} = E_e + \frac{\hbar^2}{2I} r(r + 1) + \hbar\omega_0\left(n + \frac{1}{2}\right) \tag{8.38}$$

where all symbols have been previously defined. A generic molecular transition between the initial state $E_{\text{T}}^{\text{init}}$ and the final state $E_{\text{T}}^{\text{fin}}$ is followed by the emission (molecular deexcitation) or absorption (molecular excitation) of a photon with energy equal to the difference

$$\begin{aligned}
\Delta E_{\text{T}} = E_{\text{T}}^{\text{fin}} - E_{\text{T}}^{\text{init}} &= \Delta E_e + \Delta E_{\text{rot}} + \Delta E_{\text{vib}} \\
&= \Delta E_e + \left[\frac{\hbar^2}{2I^{\text{fin}}}r^{\text{fin}}(r^{\text{fin}} + 1) - \frac{\hbar^2}{2I^{\text{init}}}r^{\text{init}}(r^{\text{init}} + 1)\right] \\
&\quad + \left[\hbar\omega_0^{\text{fin}}\left(n^{\text{fin}} + \frac{1}{2}\right) - \hbar\omega_0^{\text{init}}\left(n^{\text{init}} + \frac{1}{2}\right)\right]
\end{aligned} \tag{8.39}$$

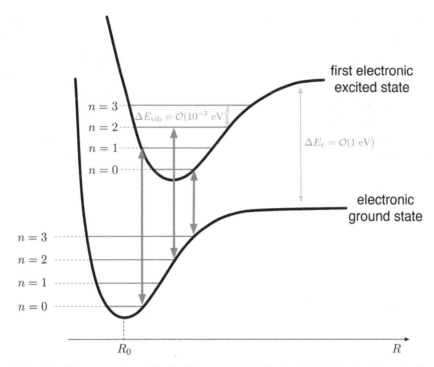

Figure 8.12. Potential energy curves (black) of the ground and first-excited electronic state and vibrational energy levels (blue) of a diatomic molecule. R is the internuclear distance. Some transitions allowed by the Franck–Condon principle are shown by arrowed red lines. In order not to make the figure too busy, rotational levels have not been drawn. The figure is not in scale.

where it has been taken into account that the moment of inertia and the fundamental vibrational frequency are different in the two molecular configurations. The frequency of the emitted/absorbed photon is $\nu = (\Delta E_e + \Delta E_{rot} + \Delta E_{vib})/h$ and typically falls in the visible or UV part of the electromagnetic spectrum.

Since $\Delta E_e \gg \Delta E_{vib} > \Delta E_{rot}$ the *molecular spectra consist in a sequence of bands.* Each element of the sequence corresponds to a given transition between two electronic states, while the fine-structure of each band is defined by all the possible roto-vibrational transitions. A very effective way to represent the situation is shown in figure 8.12.

Molecular transitions are, within the electric dipole approximation, subjected to *selection rules*; they are, however, different with respect their counterparts obtained by treating separately vibrational and rotational excitations. First of all, we observe that we can no longer disregard the fact that the potential well for nuclear vibrations is not purely harmonic: accordingly, transitions with almost any Δn can occur, as a signature of important anharmonic effects. Next, the selection rule for rotational transitions is slightly less restrictive: the condition $\Delta r = 0, \pm 1$ must be obeyed.

Transitions with $\Delta r = 0$ are now allowed[15] simply because the electronic state is also changing during the transition and, consequently, the considerations about the parity of the rotational wavefunction apply differently with respect to the discussion developed in section 7.1.1. Eventually, it is required that the total electron spin remains unaffected by the transition (therefore $\Delta S_{tot} = 0$), since exchange forces are not strong enough to provide any change.

We observe that *electronic transitions are much faster than the nuclear ones*: while the time scale for an electron to jump to another level is about 10^{-16} s, roto-vibrational transitions take of the order of 10^{-13} s. Therefore, to a good approximation we can assume that *during an electronic transition the internuclear distance remains unchanged*. This conclusion is cast in the form of the *Franck–Condon principle (first part): only vertical transitions are allowed in a diatomic molecule*. Furthermore, it must be duly taken into account that the probability $\mathcal{P}$ of electric dipole transitions is proportional to the matrix element

$$\mathcal{P} \sim \left| \int (\psi_{fin})^* \, (-e\mathbf{r}) \, \psi_{init} \, d\mathbf{r} \right|^2 \tag{8.40}$$

as obtained in section 4.2, where ψ_{fin} and ψ_{init} describe the final and initial molecular quantum state[16], respectively. It can be proved [6] that the integral appearing in equation (8.40) is proportional to the *Condon factor* $\int \psi^*_{vib,\,fin} \psi_{vib,int} \, d\mathbf{R}$, namely to the overlap between the vibrational wavefunctions of the two electronic states between which the transition occurs. This allows us to enunciate the *Franck–Condon principle (second part): the most favourite transitions occur when at least one nucleus is at the end point of its oscillations, where the vibrational wavefunction is maximum*[17]. This explains the position of the electronic vertical transitions shown in figure 8.12.

Finally, we remark that if the final electronic state corresponds to an anti-bonding molecular configuration, then the electronic transition could cause the molecule to split apart. This phenomenon is called *photochemical dissociation*

Further reading and references

[1] Levine I N 2017 *Quantum Chemistry* 7th edn (Englewood Cliffs, NJ: Prentice-Hall)

[2] Jensen F 2006 *Introduction to Computational Chemistry* 2nd edn (Hoboken, NJ: Wiley)

[3] Simons J 2003 *An Introduction to Theoretical Chemistry* (Cambridge: Cambridge University Press)

[4] Szabo A and Ostlund N S 1996 *Modern Quantum Chemistry* (New York: Dover)

[5] Demtröder W 2010 *Atoms, Molecules and Photons* (Heidelberg: Springer)

---

[15] An important remark: the transition $r^{init} = 0 \leftrightarrow r^{fin} = 0$ is in any case forbidden. Since the exchanged photon does carry an intrinsic angular momentum of $\pm\hbar$, this transition would violate the conservation of the angular momentum. A detailed explanation of this subtle effect falls beyond the scope of this manual.

[16] We mean: the wavefunction describing the full molecular state, that is, its electronic, vibrational, and rotational state.

[17] The mathematical proof of this statement is obtained by calculating the vibrational wavefunctions of a quantum oscillator moving in the anharmonic potential well provided by the $E_e(\mathbf{R})$ curve. The general solution of this problem is given in any good textbook of quantum mechanics [6, 13].

[6] Bransden B H and Joachain C J 1983 *Physics of Atoms and Molecules* (Harlow: Addison-Wesley Longman)

[7] Rigamonti A and Carretta P 2009 *Structure of Matter* 2nd edn (Milano: Springer)

[8] Atkins P and Friedman R 2011 *Molecular Quantum Mechanics* 5th edn (Oxford: Oxford University Press)

[9] Bader R F W 1995 *Atoms in Molecules: A Quantum Theory* (Cambridge: Cambridge University Press)

[10] Graham Solomons T W, Fryhle C B and Snyder S A 2013 *Organic Chemistry* 11th edn (Hoboken, NJ: Wiley)

[11] Nelson D L and Cox M 2017 *Lehninger Principles of Biochemistry* 7th edn (New York: WH Freeman)

[12] Colombo L 2021 *Solid State Physics: A Primer* (Bristol: IOP Publishing)

[13] Miller D A B 2008 *Quantum Mechanics for Scientists and Engineers* (New York: Cambridge University Press)

Part IV

Concluding remarks

IOP Publishing

Atomic and Molecular Physics (Second Edition)
A primer
Luciano Colombo

Chapter 9

What is missing in this 'Primer'

A 'Primer' cannot be, by definition, a complete and in-depth introduction to any given topic. This is especially true for such a huge field as atomic and molecular physics. This volume is no exception and, therefore, it must be acknowledged that many important issues have not been dealt with. We feel dutifully committed to mentioning them.

A concise list of missing arguments, as well as of topics which have not been thoroughly treated, is the following:

1. history of early atomic and quantum physics;
2. experimental atomic and molecular spectroscopy (including the time-resolved techniques);
3. atomic x-ray spectroscopy;
4. the principles of electron spectroscopy;
5. the physics and technology of LASER devices;
6. the physics of atom–atom and electron–atom collisions;
7. the physics of cold atoms;
8. LASER cooling and trapping techniques;
9. atom interferometry and the physics of ion traps;
10. quantum optics;
11. the fundamental concepts of quantum computing, as based on systems of trapped cold atoms;
12. the modern quantum theory of the chemical bond;
13. the *ab initio* methods for the calculation of the atomic and molecular electronic structure, together with their several different numerical implementations;
14. the physics and chemistry of organic and biological molecules.

Most of the above topics are covered by the textbooks quoted in the bibliography appended to each chapter: for further information, we refer the reader to these

doi:10.1088/978-0-7503-5734-0ch9

manuals. We observe, however, that their level is still introductory, although often more thorough than found in this Primer. The cutting-edge research in atomic and molecular physics is quite a step further: at that level, it is preferable to deal directly with the scientific literature published in specialized journals of atomic and/or molecular physics.

Part V

Appendices

IOP Publishing

Atomic and Molecular Physics (Second Edition)
A primer
Luciano Colombo

Appendix A

The formal solution of the Schrödinger equation

In this appendix we will outline the procedure to mathematically solve the Schrödinger equation for the stationary states of the hydrogen atom. A complete report on all the mathematical subtleties can be find in quantum mechanics textbooks [1–4].

Defining the mathematical problem

We have demonstrated that, by using polar coordinates under the approximations discussed in section 3.1.1, the Schrödinger equation assumes the following form

$$\frac{1}{r^2}\frac{\partial}{\partial r}\left[r^2\frac{\partial\psi(r,\theta,\phi)}{\partial r}\right] + \frac{1}{r^2\sin\theta}\frac{\partial}{\partial\theta}\left[\sin\theta\frac{\partial\psi(r,\theta,\phi)}{\partial\theta}\right] + \frac{1}{r^2\sin^2\theta}\frac{\partial^2\psi(r,\theta,\phi)}{\partial\phi^2}$$
$$= -\frac{2m_e}{\hbar^2}[E - \hat{V}(r)]\psi(r,\theta,\phi) \tag{A.1}$$

where the potential energy operator $\hat{V}(r)$ corresponds to the classical central potential

$$V(r) = -\frac{1}{4\pi\varepsilon_0}\frac{e^2}{r} \tag{A.2}$$

due to the Coulomb electron–proton coupling.

The spherical symmetry of the problem allows for the factorization

$$\psi(r,\theta,\phi) = R(r)\Theta(\theta)\Phi(\phi) \tag{A.3}$$

which, after some algebra, eventually leads to the following equation

$$\frac{\sin^2\theta}{R(r)}\frac{d}{dr}\left[r^2\frac{dR(r)}{dr}\right] + \frac{\sin\theta}{\Theta(\theta)}\frac{d}{d\theta}\left[\sin\theta\frac{d\Theta(\theta)}{d\theta}\right] + \frac{2m_e}{\hbar^2}[E - \hat{V}(r)]r^2\sin^2\theta$$
$$= -\frac{1}{\Phi(\phi)}\frac{d^2\Phi(\phi)}{d\phi^2} \tag{A.4}$$

where we have separated all terms depending on the r- and θ-variable (left hand side of the equation) from the only term containing the ϕ-variable (right hand side of the equation). Since the equation must hold for any arbitrary set of r-, θ-, and ϕ-values, we must necessarily conclude that *both sides of the equation are equal to the same constant*: this will be the starting point for the separate determination of the R-, Θ-, and Φ-functions.

Determining the Φ-functions

If we name C the constant introduced above, we can immediately set

$$\frac{d^2\Phi(\phi)}{d\phi^2} = -C\Phi(\phi) \quad \rightarrow \quad \Phi(\phi) = \Phi_0 \exp(\pm\sqrt{C}\,\phi) \tag{A.5}$$

We remark that the polar angle ϕ is defined within the interval $[0, 2\pi]$ as shown in figure 3.1. On the other hand, the function $\Phi(\phi)$ is single-valued (an imposed property discussed in section 2.1.1) or, equivalently, it must assume exactly the same value at angles ϕ and $\phi \pm 2\pi f$, where f is any natural number. This implies that $\sqrt{C} = m_l$ with m_l positive or negative integer referred to as the *magnetic quantum number*. We remark that we use the subscript l in order to stress the relation with a new quantum number l appearing soon and playing a key role in our theory. The normalization factor Φ_0 is determined by imposing $\int_0^{2\pi} |\Phi(\phi)|^2 \, d\phi = 1$ which eventually provides the explicit and complete form of the function

$$\Phi_{m_l}(\phi) = (2\pi)^{-1/2} \exp(im_l\phi) \quad \text{with } m_l = 0, \pm 1, \pm 2, \pm 3, \ldots \tag{A.6}$$

Determining the Θ-functions

Let us now focus on the left hand side term of equation (A.4) which, as already stated, is set equal to the constant $C = m_l^2$. Some further manipulation leads to

$$\frac{1}{R(r)}\frac{d}{dr}\left[r^2\frac{dR(r)}{dr}\right] + \frac{2m_e}{\hbar^2}[E - \hat{V}(r)]r^2 = -\frac{1}{\sin\theta\,\Theta(\theta)}\frac{d}{d\theta}\left[\sin\theta\frac{d\Theta(\theta)}{d\theta}\right] + \frac{m_l^2}{\sin^2\theta} \tag{A.7}$$

for which we can apply the same argument as above: both terms appearing on each side of the equation depend upon a sole variable, either r or θ, and therefore must be equal to the same constant that we name D. We begin by working on the equation for the Θ-functions

$$-\frac{1}{\sin\theta\,\Theta(\theta)}\frac{d}{d\theta}\left[\sin\theta\frac{d\Theta(\theta)}{d\theta}\right] + \frac{m_l^2}{\sin^2\theta} = D \tag{A.8}$$

The first goal is trying to define the constant D. This can be done by studying the case $m_l = 0$ which reduces to a Legendre differential equation

$$\frac{d}{d\cos\theta}\left[(1 - \cos^2\theta)\frac{d\Theta(\theta)}{d\cos\theta}\right] = -D\Theta(\theta) \tag{A.9}$$

whose solutions are Legendre polynomials

$$P_l(x) = \frac{1}{2^l l!} \frac{d^l}{dx^l} (x^2 - 1)^l \tag{A.10}$$

with $x = \cos\theta$. By direct substitution it is proved that $D = l(l+1)$ where the coefficient $l = 0, 1, 2, 3, \ldots$ is named *orbital angular momentum quantum number* (in short also referred to as: orbital quantum number). By replacing $D = l(l+1)$ into equation (A.8), the solutions for $m_l \geqslant 0$ are found to be

$$\Theta_{lm_l}(\theta) = (-1)^{m_l} \left[\frac{(2l+1)(l-m_l)!}{2(l+m_l)!} \right]^{1/2} P_l^{m_l}(\cos\theta) \tag{A.11}$$

while for $m_l < 0$ we get

$$\Theta_{lm_l}(\theta) = (-1)^{|m_l|} \Theta_{l|m_l|}(\theta) \tag{A.12}$$

In these expressions we made use of the associated Legendre polynomials

$$P_l^{m_l}(x) = (1 - x^2)^{m_l/2} \frac{d^{m_l}}{dx^{m_l}} P_l(x) \text{ with } m_l \geqslant 0 \tag{A.13}$$

The polynomial definition is mathematically meaningful only provided that $|m_l| \leqslant l$. This mathematical constraint has a very important consequence directly affecting the physics, namely: for any given integer l there are only $(2l+1)$ possible different integer values of the quantum number m_l satisfying the condition

$$-l \leqslant m_l \leqslant +l \tag{A.14}$$

We finally remark that associated Legendre polynomials are properly normalized.

Determining the *R*-functions

Let us eventually turn to the determination of the *R*-functions. The same constant $D = l(l+1)$ previously introduced must equal the left hand side of equation (A.7) leading to

$$\frac{1}{R(r)} \frac{d}{dr} \left[r^2 \frac{dR(r)}{dr} \right] + \frac{2m_e}{\hbar^2} [E - \hat{V}(r)] r^2 = l(l+1) \tag{A.15}$$

Since the quantum number l appears in the radial equation we suitably label its solutions as R_l which, for further convenience, are customarily written as

$$R_l(r) = \frac{u_l(r)}{r} \tag{A.16}$$

where $u_l(r)$ are still unknown functions satisfying the equation

$$-\frac{\hbar^2}{2m_e} \frac{d^2 u_l(r)}{dr^2} + \left[\hat{V}(r) + \frac{l(l+1)\hbar^2}{2m_e r^2} \right] u_l(r) = E u_l(r) \tag{A.17}$$

which turns out to be formally equivalent to the stationary Schrödinger equation associated with the classical situation of an electron in one-dimensional motion subjected to the effective potential

$$W = -\frac{e^2}{4\pi\varepsilon_0}\frac{1}{r} + \frac{l(l+1)\hbar^2}{2m_e}\frac{1}{r^2} \tag{A.18}$$

This corresponds to a well-known quantum mechanical problem whose solution provides a recursive expression for $u_l(r)$. The normalized radial functions are finally obtained for any state with $E < 0$ (corresponding to the situation of a bound electron–proton pair) as

$$R_{nl}(r) = -\sqrt{\left(\frac{2}{na_0}\right)^3 \frac{(n-l-1)!}{2n[(n+l)!]^3}} \, \exp\left(-\frac{r}{na_0}\right)\left(\frac{2r}{na_0}\right)^l L_{n+l}^{2l+1}\left(\frac{2r}{na_0}\right) \tag{A.19}$$

where $L_q^p(x)$ are the associated Laguerre polynomials

$$L_q^p(x) = \frac{d^p}{dx^p}\left\{\exp(x)\frac{d^q}{dx^q}[x^q \exp(-x)]\right\} \tag{A.20}$$

and n is named *principal quantum number*. A constraint emerging from the mathematics is that

$$l \leqslant n - 1 \tag{A.21}$$

or, equivalently, we can state that *for any assigned n the only possible values of the angular quantum number are l = 0, 1, 2, ... , n − 1*. We finally remark that at this stage the quantity a_0 appearing in equation (A.19) is just an abbreviation for

$$a_0 = \frac{\varepsilon_0 h^2}{\pi m_e e^2} \tag{A.22}$$

Interestingly enough, this quantity corresponds to the Bohr radius, previously introduced on the basis of purely phenomenological arguments in equation (1.22). The physics behind this unexpected coincidence is thoroughly discussed in chapter 3.

Further reading and references

[1] Bransden B H and Joachain C J 1983 *Physics of Atoms and Molecules* (Harlow: Addison-Wesley Longman)
[2] Miller D A B 2008 *Quantum Mechanics for Scientists and Engineers* (New York: Cambridge University Press)
[3] Morrison M A, Estle T L and Lane N F 1976 *Quantum States of Atoms, Molecules, and Solids* (Englewood Cliffs, NJ: Prentice-Hall)
[4] Sakurai J J and Napolitano J 2011 *Modern Quantum Mechanics* 2nd edn (Reading, MA: Addison-Wesley)

IOP Publishing

Atomic and Molecular Physics (Second Edition)
A primer
Luciano Colombo

Appendix B

The spin–orbit interaction energy in hydrogenic atoms

In section 3.4 it has been proved that the spin–orbit interaction energy in a hydrogenic atom with nuclear charge $+Ze$ is

$$E_{so} = \xi(r)\, \mathbf{S} \cdot \mathbf{L} \tag{B.1}$$

where

$$\xi(r) = \frac{1}{2m_e^2 c^2}\, \frac{1}{r}\, \frac{dV(r)}{dr} \tag{B.2}$$

is spin–orbit coupling coefficient, while

$$V(r) = -\frac{1}{4\pi\varepsilon_0}\, \frac{Ze^2}{r} \tag{B.3}$$

is the nuclear Coulomb potential energy. In order to evaluate quantitatively E_{so} we need to calculate the quantum mechanical expectation value of the coupling coefficient $\xi(r)$.

Since we know that E_{so} is rather small (see discussion in section 3.4), we can assume that the hydrogenic wavefunctions are still a valid description of the electron quantum states. We remark that $\xi(r)$ does not contain the spin coordinate and it only depends on the r observable: accordingly, the corresponding quantum operator is just formally identical to the expression given in equation (B.2). Its expectation value is calculated according to equation (2.5), i.e., without using the spin part of the wavefunctions

$$\langle \xi(r) \rangle = \frac{1}{2m_e^2 c^2}\, \frac{Ze^2}{4\pi\varepsilon_0}\, \left\langle \frac{1}{r^3} \right\rangle \tag{B.4}$$

The latter term appearing in the right-hand-side of this equation is tabulated in any quantum mechanics textbook

doi:10.1088/978-0-7503-5734-0ch11

$$\left\langle \frac{1}{r^3} \right\rangle = \frac{Z^3}{a_0^3 \, n^3 \, l\left(l + \frac{1}{2}\right)(l + 1)} \tag{B.5}$$

where we assumed infinite nuclear mass. Eventually we get

$$\langle \xi(r) \rangle = \frac{1}{2m_e^2 c^2} \left(\frac{e^2}{4\pi\varepsilon_0} \right) \frac{Z^4}{a_0^3} \frac{1}{n^3 \, l\left(l + \frac{1}{2}\right)(l + 1)} \tag{B.6}$$

Appendix C

The hyperfine structure in hydrogenic atoms

Nuclear physics provides robust evidence that nuclei are not only electrically charged, but they also carry a magnetic moment [1]. This characteristic is described by defining the *nuclear magnetic moment* $\mathbf{M}_N$ as

$$\mathbf{M}_N = g_N \frac{\mu_N}{\hbar} \mathbf{N} \tag{C.1}$$

where g_N is a dimensionless constant named *nuclear g-factor*, μ_N is the *nuclear magneton*

$$\mu_N = \frac{m_e}{m_p} \mu_B = 5.050\,82 \times 10^{-27} \text{ J T}^{-1} \tag{C.2}$$

defined as the *nuclear Bohr magneton* (m_p is the proton mass), and $\mathbf{N}$ is the *nuclear spin* or, equivalently, the total nuclear angular momentum.

We start by remarking that $\mu_N \ll \mu_B$ and, therefore, nuclear magnetic effects are expected to be really very small, actually smaller than the fine structure ones: they can treated as perturbations on the hydrogenic quantum states. Next, we observe that *the nuclear magnetic dipole interacts with the magnetic field* $\mathbf{B}_e$ *generated by the electron*. Its origin is twofold[1]: it derives (i) from the electron orbital motion (generating a magnetic field that can be calculated similarly to the case of the spin–orbit interaction), and (ii) from the dipole–dipole interaction between the electron and nuclear spins. In summary, a classical hyperfine energy term E_{hf} is added to the atomic system, which is cast in the form

$$E_{hf} = -\mathbf{M}_N \cdot \mathbf{B}_e \tag{C.3}$$

In order to evaluate such an energy, the same approach used for the spin–orbit interaction is followed: a classical analysis based on the 'vector model' and leading to a full expression of E_{hf} is at first evaluated; next, its quantum-mechanical

---

[1] It also includes the so-called Fermi contact interaction which has no classical counterpart and it is important for *s*-states for which the electron has a non-vanishing probability to be near the nucleus.

expectation value is calculated on the basis of the hydrogenic wavefunctions solutions of equation (3.11).

For illustration purposes, we only focus on the major contribution due the interaction of the nuclear magnetic moment with the electron orbital current. By introducing the *total atomic angular momentum* $\mathbf{F} = \mathbf{J} + \mathbf{N}$, it is proved that each fine structure level E_{nj} of the hydrogen atom provided in equation (3.95) is corrected as

$$E_{njF} = E_{nj} + \alpha_{\text{hf}}[F(F + 1) - j(j + 1) - N(N + 1)] \tag{C.4}$$

where F and N are the quantum numbers associated with the total electronic angular momentum and with the nuclear angular momentum, respectively, and

$$\alpha_{\text{hf}} = \frac{g_N \mu_N}{2\sqrt{j(j + 1)}} B_{\text{e},j} \tag{C.5}$$

is the *hyperfine splitting constant*, with $B_{\text{e},j}$ the modulus of the magnetic field felt by the nucleus because of the orbiting electron in the state described by the total angular momentum quantum number j. This equation defines the *hyperfine energy spectrum of hydrogen*.

We conclude by qualitatively estimating the hyperfine effects on the hydrogen ground state. Since it is known experimentally that $N = 1/2$ and we know that for such a state $j = 1/2$, then we easily get two possible values for the quantum number F, namely $F = 0$ and $F = 1$. This is tantamount to saying that *the hydrogen ground state is split into a doublet*. The corresponding separation energy is measured to be of the order of $\Delta E_{\text{hf}} \sim 6 \times 10^{-6}$ eV, proving that these effects are indeed very small.

Actually, there exist more hyperfine effects than those merely related to the nuclear spin. Other hyperfine effects are referred to as *isotope shifts* since they do not generate splittings but, rather, just shifts of the energy levels. Basically, they are due to either mass or volume effects: they are caused by the fact that the nuclear mass, although much larger than the electron one, is not infinite and by the fact that its charge is distributed over a finite volume. It was quite reasonable to assume that nuclei are point-like charges since their dimension is much smaller than the typical electron–nucleus distances. In fact, they are not really point-like and, in general, the nuclear charge is not spherically distributed: *nuclear electric quadrupole moment* effects should be duly included in the ultimate description of the energy spectrum.

The full treatment of any hyperfine effect is thoroughly discussed in [2–4]. Overall, they produce the energy spectrum reported in figure 3.17 for the hydrogen atom.

Further reading and references

[1] Povh B, Rith K, Scholz C and Zetsche F 2009 *Particles and Nuclei* 6th edn (Heidelberg: Springer)
[2] Bransden B H and Joachain C J 1983 *Physics of Atoms and Molecules* (Harlow: Addison-Wesley Longman)
[3] Demtröder W 2010 *Atoms, Molecules and Photons* (Heidelberg: Springer)
[4] Foot C J 2005 *Atomic Physics* (Oxford: Oxford University Press)

Appendix D

Nucleus finite-size correction

In order to implement the notion that atomic nuclei have a finite size [1], we assume that the nuclear charge $+Ze$ is uniformly distributed inside a sphere of radius r_N. This certainly corresponds to a simplified picture which, nevertheless, contains the essential elements enabling us to evaluate the nucleus finite-size correction to the atomic spectrum of hydrogenic atoms.

Maxwell electromagnetism [2, 3] predicts the electrostatic interaction energy between the electron charge and the nuclear sphere to be

$$V(r) = \begin{cases} -\dfrac{1}{4\pi\varepsilon_0}\dfrac{Ze^2}{r} & r \geqslant r_N \\[2mm] \dfrac{Ze^2}{4\pi\varepsilon_0}\dfrac{1}{2r_N}\left[\left(\dfrac{r}{r_N}\right)^2 - 3\right] & r < r_N \end{cases} \tag{D.1}$$

which represents a correction to the model in which the nucleus is considered to be point-like. We treat this correction as a perturbation to the electron energy which we describe quantum mechanically by the following operator

$$\hat{W}(r) = \begin{cases} 0 & r \geqslant r_N \\[2mm] \dfrac{Ze^2}{4\pi\varepsilon_0}\dfrac{1}{2r_N}\left[\left(\dfrac{r}{r_N}\right)^2 - 3 + 2\dfrac{r_N}{r}\right] & r < r_N \end{cases} \tag{D.2}$$

where we are making use of the formalism developed in section 2.7.

We now focus our attention on the ground state of an hydrogenic atom described by the wavefunction

$$\psi_{100}(r) = 2\sqrt{\frac{Z^3}{4\pi a_0^3}}\,\exp[-Zr/a_0] \tag{D.3}$$

doi:10.1088/978-0-7503-5734-0ch13

as discussed in section 3.2. By applying the time-independent perturbation theory, we calculate the first-order correction ΔE_1 to the ground state energy $E_1 = -13.6$ eV as

$$\Delta E_1 = \int \psi_{100}^* \hat{W} \, \psi_{100} \, d\mathbf{r}$$

$$= \frac{1}{4\pi\varepsilon_0} \frac{2Z^4 e^2}{r_N a_0^3} \int_0^{r_N} r^2 \left[\left(\frac{r}{r_N} \right)^2 - 3 + 2\frac{r_N}{r} \right] \exp[-2Zr/a_0] \, d\mathbf{r} \tag{D.4}$$

which is easily calculated by using integral tables [4]. The calculation is quite laborious, but it eventually leads to the following result

$$\Delta E_1 = \frac{2}{5} \frac{1}{4\pi\varepsilon_0} \frac{Z^4 e^2}{a_0} \left(\frac{r_N}{a_0} \right)^2 \tag{D.5}$$

which indicates that the correction is weighted by a factor $(r_N/a_0)^2 \sim 10^{-8}$ and, therefore, for the hydrogen atom ($Z = 1$) it is indeed very small. We also remark that it is anyway the largest correction, since the electron in the ground $1s$ state has the higher probability to approach the nuclear region (see figure 3.4). On the other hand, the nucleus finite-size correction can be relevant for hydrogenic atoms with large enough Z; in this case, either the Z^4 prefactor compensates the (r_N/a_0) term and the Bohr radius a_0 must be replaced by its smaller effective counterpart provided in equation (3.39), thus bringing the value of the (r_N/a_0) closer to 1.

Further reading and references

[1] Povh B, Rith K, Scholz C and Zetsche F 2009 *Particles and Nuclei* 6th edn (Heidelberg: Springer)
[2] Jackson J D 1975 *Classical Electromagnetism* (New York: Wiley)
[3] Feynman R P, Leighton R B and Sands M 1963 *The Feynman Lectures on Physics* (Reading, MA: Addison-Wesley)
[4] Gradshteyn I S and Ryzhik I M 1980 *Tables of Integrals Series, and Products* 8th edn (San Diego, CA: Academic)

IOP Publishing

Atomic and Molecular Physics (Second Edition)
A primer
Luciano Colombo

Appendix E

Boltzmann statistics

Let us consider N identical and structureless classical[1] particles, occupying states with discrete energy E_i with $i = 1, 2, 3, \ldots$ and $E_1 < E_2 < E_3 < \cdots$ so that their total number and the system internal energy $\mathcal{U}$ are, respectively, written as

$$N = \sum_i n_i \text{ and } \mathcal{U} = \sum_i n_i E_i \qquad (E.1)$$

if n_i is the number of particles with energy E_i.

For an isolated system the internal energy $\mathcal{U}$ is conserved. However, collisions and interactions affect the *distribution of the particles among the energy states* which is referred to as *partition*; this implies that the occupation numbers n_i may change in time and the partition (or, equivalently, the system *microstate*) may accordingly change. A natural question arises at this point: *which is the most probable partition, once the macroscopic thermodynamical parameters of the system (volume, number of particles, and internal energy) have been assigned?* It is of fundamental importance to answer this question, since *a system in its most probable partition is said to be at equilibrium*. This statement can be easily justified since it is quite obvious to assume that the free evolution of an isolated system is always from a partition with lower probability to a partition with higher probability.

In order to work out the equilibrium partition, we will for the moment consider the particles forming the system to be identical but nevertheless *distinguishable*. Let us further assume that all their allowed states with energy E_i have the *same probability of being occupied*. It is important to remark that both these assumptions and the distinguishability requirement will be soon removed, in order to elaborate a more general picture; at this stage they must be intended just as pedagogical tools. Any given partition can be realised in different ways, each corresponding to a specific distribution of the particles among the states. It is quite realistic to consider that *the probability of each partition is proportional to the number of different ways it*

---

[1] That is, spinless particles.

can be realised. We will further base our arguments on the quite reasonable statement that *the probability of each partition is proportional to the number of different ways it can be realised*. It is therefore mandatory to correctly count the number of different ways we can place $n_i \leqslant N$ particles on the state with energy E_i, once the remaining particles have been already placed.

If we start from the lowest level E_1 such a number is $N!/n_1!(N - n_1)!$, corresponding to the number of permutations of N objects taken n_1 at a time. For the second state with energy E_2 the number of available particles to accomodate is $(N - n_1)$ and, therefore, it is easy to understand that $(N - n_1)!/n_2!(N - n_1 - n_2)!$ is the number of different ways we can place n_2 particles on it. By iterating this procedure for all the remaining states, we eventually obtain *the number of different distinguishable ways a partition can be realised with n_1 particles on the state E_1, n_2 on the state E_2, n_3 on the state E_3, and so on*

$$\frac{N!}{n_1!n_2!n_3!\dots} \tag{E.2}$$

while, as stated above, the corresponding probability P for this partition is proportional to it.

We now proceed by refining and generalizing our reasoning and, first of all, we admit that the allowed states have different occupation probabilities which we will indicate with p_i (in other words, p_i is the intrinsic probability of finding one particle with energy E_i). This implies that the number of distinguishable ways we can realise the above partition must be corrected as

$$\frac{N!p_1^{n_1}p_2^{n_2}p_3^{n_3}\dots}{n_1!n_2!n_3!\dots} \tag{E.3}$$

since the probability of finding n_i particles on the state E_i is precisely $p_i^{n_i}$. It is very important to remark that *we are not applying any restriction on such intrinsic (or state) probabilities p_i*. In other words, we are not assuming any exclusion principle related to the character of the total wavefunction. This is tantamount to stating that *we are developing a purely classical statistical theory*.

Next, we remove the indistinguishability assumption (in other words, particle labelling is no longer relevant): this implies that the $N!$ permutations among particle pairs that occupy the different states actually provide the very same partition. Therefore, *the number Ω of different ways the partition can be realised* is eventually corrected in

$$\Omega = \frac{1}{N!}\frac{N!p_1^{n_1}p_2^{n_2}p_3^{n_3}\dots}{n_1!n_2!n_3!\dots} = \frac{p_1^{n_1}p_2^{n_2}p_3^{n_3}\dots}{n_1!n_2!n_3!\dots} = \prod_i \frac{p_i^{n_i}}{n_i!} \tag{E.4}$$

which corresponds to a probability

$$P = \xi \prod_i \frac{p_i^{n_i}}{n_i!} \tag{E.5}$$

where ξ is a convenient proportionality factor linking the number of distinguishable ways the partition can be realised to its corresponding probability P. This factor does not play any role in the theory, as will be clear very soon.

Since, as stated above, the equilibrium state corresponds to the most probable partition, it can be calculated by maximizing the quantity provided in equation (E.5). This procedure, however, must fulfil two physical constraints provided in equations (E.1), namely (i) the conservation of energy and (ii) the conservation of the number of particles. In practice, the calculation proceeds by considering at first the natural logarithm[2] of the partition probability P

$$\ln P = \ln \xi + \ln \left(\prod_i \frac{p_i^{n_i}}{n_i!} \right) = \ln \xi + \sum_i \ln \frac{p_i^{n_i}}{n_i!} = \ln \xi + \sum_i (n_i \ln p_i - \ln n_i!) \quad \text{(E.6)}$$

which is easily calculated by means of the Stirling formula $\ln n! \simeq n \ln n - n$ valid for large enough n numbers

$$\ln P = \ln \xi + \sum_i \left(-n_i \ln \frac{n_i}{p_i} + n_i \right) = \ln \xi + N - \sum_i n_i \ln \frac{n_i}{p_i} \quad \text{(E.7)}$$

Next, the maximum probability condition is mathematically set as $-d \ln P = 0$, where the differential variation is calculated with respect to small changes $dn_1, dn_2, dn_3,...$ in the occupation numbers

$$d \ln P = -\sum_i dn_i \ln \frac{n_i}{p_i} = 0 \quad \text{(E.8)}$$

where we have taken into consideration that both ξ and N are constant since, respectively, the first one is just a numerical factor and the system is isolated. The two constraints, however, impose that upon the same small changes we must have

$$\sum_i dn_i = 0 = \sum_i E_i \, dn_i \quad \text{(E.9)}$$

so that the maximum condition provided in equation (E.8) must be compensated according to the Lagrange method

$$\sum_i \left(\ln \frac{n_i}{p_i} + \alpha + \beta E_i \right) dn_i = 0 \quad \text{(E.10)}$$

where α and β are the Lagrange multipliers. The maximum probability is then found by imposing

$$\ln \frac{n_i}{p_i} + \alpha + \beta E_i = 0 \quad \text{(E.11)}$$

---

[2] Since the logarithm is a monotonic function of its variable, the maximum of $\ln P$ corresponds to the maximum of P.

which leads to

$$n_i = p_i \exp(-\alpha - \beta E_i) \tag{E.12}$$

the fundamental result paving the way to the statistical mechanics of classical identical particles.

In order to proceed further we need to determine the actual value of the Lagrange multipliers α and β appearing in equation (E.12). To this aim, let us calculate explicitly the total number of particles as

$$N = \sum_i n_i = \sum_i p_i \exp(-\alpha - \beta E_i) = \exp(-\alpha) \sum_i p_i \exp(-\beta E_i) \tag{E.13}$$

and introduce the *partition function* $\mathcal{Z}$

$$\mathcal{Z} = \sum_i p_i \exp(-\beta E_i) \tag{E.14}$$

which represents a key quantity of statistical physics [1]. By combining equations (E.13) and (E.14) we obtain $\exp(-\alpha) = N/\mathcal{Z}$ and eventually write

$$n_i = \frac{N}{\mathcal{Z}} p_i \exp(-\beta E_i) \tag{E.15}$$

which is known as *the Boltzmann distribution law* describing the equilibrium statistics of an assembly of classical identical particles distributed on energy levels E_i with intrinsic probability p_i. We remark that *the parameter β is necessarily linked to the notion of temperature*, since we are treating an equilibrium state which is naturally associated with a given value of temperature. For historical reasons it has been preferred to introduce a new physical quantity T, referred to as *absolute temperature*, defined as

$$k_B T = \frac{1}{\beta} \tag{E.16}$$

where the Boltzmann constant k_B allows us to express $1/\beta$ in energy units.

Further reading and references

[1] Colombo L 2022 *Statistical Physics of Condensed Matter Systems: A primer* (Bristol: IOP Publishing)

IOP Publishing

Atomic and Molecular Physics (Second Edition)
A primer
Luciano Colombo

Appendix F

The 'minimal coupling' scheme

Let us consider a particle with mass m and charge q under the action of an electromagnetic field described by the electrostatic potential V and by the vector field $\mathbf{A}$. If the particle moves with velocity v, it experiences the Lorentz force

$$\mathbf{F} = q(\mathbf{E} + v \times \mathbf{B}) \tag{F.1}$$

which is derived by a generalized potential

$$U = q(V - v \cdot \mathbf{A}) \tag{F.2}$$

The corresponding Lagrangian function $\mathcal{L}$ for the particle is

$$\mathcal{L} = \frac{1}{2}mv^2 - qV + q\,v \cdot \mathbf{A} \tag{F.3}$$

from which we obtain

$$\mathbf{P} = mv + q\mathbf{A} \tag{F.4}$$

representing the particle *generalized momentum* $\mathbf{P}$. From this result we immediately derive the ordinary particle momentum $\mathbf{p} = mv = \mathbf{P} - q\mathbf{A}$ and, therefore, the classical Hamiltonian function

$$H = \frac{1}{2m}\,(\mathbf{P} - q\mathbf{A})^2 + qV \tag{F.5}$$

so that the corresponding quantum operator can be easily found in the form reported in equation (4.17).

To this aim, it is instructive to develop the kinetic energy term appearing in equation (F.5)

$$\frac{1}{2m}\,(\mathbf{P} - q\mathbf{A})^2 = \frac{1}{2m}\,(P^2 - 2q\,\mathbf{P} \cdot \mathbf{A} + q^2\,A^2)$$
$$= \frac{P^2}{2m} - \frac{q}{m}\mathbf{P} \cdot \mathbf{A} + \frac{q^2A^2}{2m} \tag{F.6}$$

doi:10.1088/978-0-7503-5734-0ch15

where we immediately recognise that, under the assumption of weak electromagnetic field, the quadratic term $q^2 A^2 / 2m$ can be considered negligibly small: it will be hereafter neglected. Accordingly, we elaborate the correspondence

$$\underbrace{\frac{1}{2m} (\mathbf{P} - q\mathbf{A})^2}_{\text{classical}} \rightarrow \underbrace{-\frac{\hbar^2}{2m} \nabla^2 + \frac{iq\hbar}{2m} (\nabla \cdot \mathbf{A} + \mathbf{A} \cdot \nabla)}_{\text{quantum}} \tag{F.7}$$

providing the quantum operator associated with the classical energy term reported on the left and side of equation (F.6).

In order to further proceed, we make use of the commuting rule $\nabla \cdot \mathbf{A} - \mathbf{A} \cdot \nabla = 0$ obtained by inserting the Coulomb gauge condition $\nabla \cdot \mathbf{A} = 0$ into the following development

$$\nabla \cdot (\mathbf{A}\psi) = (\nabla \cdot \mathbf{A})\psi + \mathbf{A} \cdot (\nabla\psi) = \mathbf{A} \cdot (\nabla\psi) \tag{F.8}$$

defining the operator action on a generic wavefunction. Eventually, we get

$$\underbrace{\frac{1}{2m} (\mathbf{P} - q\mathbf{A})^2}_{\text{classical}} \rightarrow \underbrace{-\frac{\hbar^2}{2m} \nabla^2 + \frac{iq\hbar}{m} \mathbf{A} \cdot \nabla}_{\text{quantum}} \tag{F.9}$$

which by setting $q = -e$ and $m = m_e$ corresponds to equation (4.18).

Atomic and Molecular Physics (Second Edition)
A primer
Luciano Colombo

Appendix G

Screening effects in He atom

Let us consider the ground state of the He atom. The interaction between the nuclear charge and each electron is screened by the presence of the second electron: we will describe this effect by replacing the bare nuclear charge $+Ze$ with a screened one $+Z^*e$. Accordingly, within the independent electron approximation, the total electron wavefunction Ψ_{GS} for the ground state can be written as a product of two $1s$ hydrogenic single-particle wavefunctions (see tables 3.1 and 3.2, as well as section 3.3)

$$\Psi_{GS} = \frac{1}{\pi}\left(\frac{Z^*}{a_0}\right)^3 \exp[-Z^*(r_1 + r_2)/a_0] \tag{G.1}$$

where we have made use of the screened nuclear charge value. This trial function allows us to predict the ground state energy as

$$E_{GS}(Z^*) = \int \psi_{GS}^* \left[-\frac{\hbar^2}{2m_e}(\nabla_1^2 + \nabla_2^2) - \frac{1}{4\pi\varepsilon_0}\frac{Ze^2}{r_1} - \frac{1}{4\pi\varepsilon_0}\frac{Ze^2}{r_2} + \frac{1}{4\pi\varepsilon_0}\frac{e^2}{r_{12}} \right] \psi_{GS} \, d\mathbf{r_1}d\mathbf{r_2}$$

$$= \underbrace{-\frac{\hbar^2}{2m_e} \int \psi_{GS}^*(\nabla_1^2 + \nabla_2^2)\psi_{GS} \, d\mathbf{r_1}d\mathbf{r_2}}_{A} + \tag{G.2}$$

$$+ \underbrace{\frac{1}{4\pi\varepsilon_0} \int \psi_{GS}^*\left(\frac{e^2}{r_{12}}\right)\psi_{GS} \, d\mathbf{r_1}d\mathbf{r_2}}_{B} \underbrace{- \frac{Ze^2}{4\pi\varepsilon_0} \int \psi_{GS}^*\left(\frac{1}{r_1} + \frac{1}{r_2}\right)\psi_{GS} \, d\mathbf{r_1}d\mathbf{r_2}}_{C}$$

where '1' and '2' labels are defined in figure 5.2. The three terms A, B, and C can be calculated separately.

The A term is just twice the kinetic energy E_{kin} of an electron subjected to a Coulomb field generated by a charge $+Z^*e$, namely: $A = 2E_{kin}$. Since for the virial theorem we have $E_{kin} = -V/2$ where V is the Coulomb potential energy, then we can write

$$A = -V = -\left[-\frac{(Z^*e)^2}{4\pi\varepsilon_0 a_0} \right] = 2(Z^*)^2|E_{1s}^{(H)}| \tag{G.3}$$

doi:10.1088/978-0-7503-5734-0ch16

where $|E_{1s}^{(H)}| = 13.6\,\text{eV}$ is the absolute value of the ground state energy of the hydrogen atom.

The B term is calculated straightforwardly

$$B = \frac{5}{8}\,Z^* \,\frac{e^2}{a_0} = \frac{5}{4}\,Z^* \,|E_{1s}^{(H)}| \tag{G.4}$$

Finally, in order to evaluate the C term we must use the expectation value Z^*/a_0 for the $1/r$ operator calculated on the trial wavefunction given in equation (G.1). We obtain

$$C = -\frac{Ze^2}{4\pi\varepsilon_0}\,2\frac{Z^*}{a_0} = -4ZZ^* \,|E_{1s}^{(H)}| \tag{G.5}$$

and by summing the above three results we get

$$E_{GS}(Z^*) = A + B + C = 2(Z^*)^2|E_{1s}^{(H)}| + \frac{5}{4}\,Z^* \,|E_{1s}^{(H)}| - 4ZZ^* \,|E_{1s}^{(H)}| \tag{G.6}$$

Since this is the ground state energy it must hold

$$\frac{dE_{GS}(Z^*)}{dZ^*} = 0 = 4Z^* + \frac{5}{4} - 4Z \tag{G.7}$$

from which we eventually obtain the value of the screened nuclear charge

$$Z^* = Z - \frac{5}{16} \tag{G.8}$$

so that we can write

$$E_{GS} = -\left(2Z^2 - \frac{5}{4}Z + \frac{25}{128}\right)|E_{1s}^{(H)}| \tag{G.9}$$

as the corrected energy for the ground state.

IOP Publishing

Atomic and Molecular Physics (Second Edition)
A primer
Luciano Colombo

Appendix H

The Thomas–Fermi method

The basic idea of the semi-classical Thomas–Fermi method is to describe the N electrons of a neutral atom in its ground state as a *free electron gas*. The gas is supposed to be confined in the atomic region by a radial potential $V_{TF}(r)$, hereafter referred to as the *Thomas–Fermi potential*. While this approximation is clearly very strong, it works surprisingly well at least for atoms with a large N.

Elementary quantum mechanics [1] proves that at $T = 0$ K the energy E_F (known as Fermi energy) of the highest occupied state of a free electron gas is

$$E_F = \frac{\hbar^2}{2m_e}(3\pi^2\rho)^{2/3} \tag{H.1}$$

where ρ is its uniform density. By exporting this result to the case of our current interest, we can relate the confining potential $V_{TF}(r)$ to the atomic electron density through the Poisson equation

$$\nabla^2 V_{TF}(r) = -\frac{\rho(r)}{\varepsilon_0} \tag{H.2}$$

where by writing $\rho = \rho(r)$ we made explicit the notion that, at variance with the homogeneous free electron gas, the atom must be modelled by a density with some radial dependence. Another relation between the potential and the density can be derived by setting the maximum energy E_{max} of an electron in the gas as

$$E_{max} = E_F + V_{TF}(r) \tag{H.3}$$

leading to

$$\rho(r) = \frac{1}{3\pi^2}\left(\frac{2m_e}{\hbar^2}\right)^{3/2}[E_{max} - V_{TF}(r)]^{3/2} \tag{H.4}$$

We remark that this results is obtained by taking into account that the Thomas–Fermi potential must obey two boundary conditions, namely: (i) for $r \to 0$ we have $V_{TF}(r) \to -Ze^2/4\pi\varepsilon_0 r$, while (ii) for $r \to \infty$ we have $V_{TF}(r) \to 0$ (this condition, in

doi:10.1088/978-0-7503-5734-0ch17

particular, allows us to set to zero a constant term in the potential). Furthermore, the following condition

$$\int_0^{r_0} \rho(r)\, 4\pi r^2 dr = N \tag{H.5}$$

must be imposed for the atomic electron system, which is confined within a sphere of radius r_0 representing the 'atomic volume'.

By combining equation (H.2) and (H.4) we get the *Thomas–Fermi equation*

$$\frac{1}{r}\frac{d^2}{dr^2}[r\, V_{\text{TF}}(r)] = \frac{e^2}{3\pi^2 \varepsilon_0}\left(\frac{2m_{\text{e}}}{\hbar}\right)^{3/2}[V_{\text{TF}}(r)]^{3/2} \tag{H.6}$$

which is a universal second-order, non-linear equation with no analytical solution.

The numerical solution of equation (H.6) [2] provides either a rough estimate of the central-field to be used in direct applications (like for instance in the calculation of the atomic form factors used to predict the scattering of x-rays by an atom) or a first guess for the potential to be determined by a self-consistent procedure.

Further reading and references

[1] Colombo L 2022 *Statistical Physics of Condensed Matter Physics: A primer* (Bristol: IOP Publishing)
[2] Bransden B H and Joachain C J 1983 *Physics of Atoms and Molecules* (Harlow: Addison-Wesley Longman)

Atomic and Molecular Physics (Second Edition)
A primer
Luciano Colombo

Appendix I

The formal proof of the Born–Oppenheimer approximation

We provide a simple derivation[1] of equations (6.6) and (6.7). In order to make formalism as lean as possible, we will make use of the compact notation $\psi_n(\mathbf{R}) = \psi_n$ and $\psi_e^{(\mathbf{R})}(\mathbf{r}) = \psi_e$.

By inserting equation (6.5) into equation (6.4) we have

$$\left[-\frac{\hbar^2}{2}\sum_\alpha \frac{1}{M_\alpha}\nabla_\alpha^2 - \frac{\hbar^2}{2m_e}\sum_i \nabla_i^2 + \hat{V}_{ne} + \hat{V}_{nn} + \hat{V}_{ee} \right]\psi_n\psi_e = E_T\psi_n\psi_e \tag{I.1}$$

Let us consider at first the action of the kinetic energy operators

$$\begin{aligned} \nabla_\alpha^2(\psi_n\psi_e) &= \psi_e\nabla_\alpha^2\psi_n + \psi_n\nabla_\alpha^2\psi_e + 2\nabla_\alpha\psi_e \cdot \nabla_\alpha\psi_n \\ \nabla_i^2(\psi_n\psi_e) &= \psi_n\nabla_i^2\psi_e \end{aligned} \tag{I.2}$$

where we adopted the standard procedure

$$\begin{aligned} \nabla_\alpha^2(\psi_n\psi_e) &= \nabla_\alpha \cdot \nabla_\alpha(\psi_n\psi_e) \\ &= \nabla_\alpha \cdot (\psi_n\nabla_\alpha\psi_e + \psi_e\nabla_\alpha\psi_n) \\ &= \psi_e\nabla_\alpha^2\psi_n + \psi_n\nabla_\alpha^2\psi_e + 2\nabla_\alpha\psi_e \cdot \nabla_\alpha\psi_n \end{aligned} \tag{I.3}$$

Let us now re-organize equation (I.1) more conveniently

$$\begin{aligned} &\left[-\hbar^2\sum_\alpha\left(\frac{1}{M_\alpha}\nabla_\alpha\psi_e \cdot \nabla_\alpha\psi_n + \frac{1}{2M_\alpha}\psi_n\nabla_\alpha^2\psi_e \right) \right] \\ &\quad - \frac{\hbar^2}{2}\psi_e\sum_\alpha\frac{1}{M_\alpha}\nabla_\alpha^2\psi_n - \frac{\hbar^2}{2m_e}\psi_n\sum_i\nabla_i^2\psi_e \\ &\quad + (\hat{V}_{nn} + \hat{V}_{ne} + \hat{V}_{ee})\psi_n\psi_e = E_T\psi_n\psi_e \end{aligned} \tag{I.4}$$

---

[1] While the formal development here presented is tailored to match the level of this manual, a more complete proof of the adiabatic approximation is found in [1–4]. We recall that Latin and Greek labels indicate electron and nuclear coordinates, respectively.

The Born–Oppenheimer approximation consists in *neglecting all the mixed terms* appearing in the first row of this equation, a choice which allows us to reformulate equation (I.1) in the simplified form

$$-\frac{\hbar^2}{2}\psi_e\sum_\alpha\frac{1}{M_\alpha}\nabla_\alpha^2\psi_n + \psi_n\left[-\frac{\hbar^2}{2m_e}\sum_i\nabla_i^2\psi_e + (\hat{V}_{ee} + \hat{V}_{ne})\psi_e\right] + \hat{V}_{nn}\psi_n\psi_e = E_T\psi_n\psi_e \quad (I.5)$$

We now observe that the term in square parenthesis actually corresponds to the *eigenvalue problem for the electrons* anticipated in equation (6.6)

$$\left[-\frac{\hbar^2}{2m_e}\sum_i\nabla_i^2\psi_e + (\hat{V}_{ee} + \hat{V}_{ne})\psi_e\right] = E_e^{(R)}\psi_e \quad (I.6)$$

By inserting this result into equation (I.5) we get

$$-\frac{\hbar^2}{2}\psi_e\sum_\alpha\frac{1}{M_\alpha}\nabla_\alpha^2\psi_n + \psi_e\psi_nE_e^{(R)} + \psi_e\hat{V}_{nn}\psi_n = E_T\psi_n\psi_e \quad (I.7)$$

which immediately leads to

$$\left[-\frac{\hbar^2}{2}\sum_\alpha\frac{1}{M_\alpha}\nabla_\alpha^2 + \hat{V}_{nn}\right]\psi_n = [E_T - E_e^{(R)}]\psi_n \quad (I.8)$$

This equation corresponds to the *eigenvalue equation for the nuclear problem* anticipated in equation (6.7).

While our goal to provide a formal proof of equations (6.6) and (6.7) has been reached, we still need to provide a convincing quantitative argument in support of the adopted Born–Oppenheimer approximation. In order to do so, we calculate the quantum expectation value of any disregarded term for a generic state whose total molecular wavefunction is $\Psi = \psi_n\psi_e$.

The expectation value of the first neglected mixed term is proportional to

$$\int \psi_e^*\nabla_\alpha\psi_e d\mathbf{r} = \frac{1}{2}\nabla_\alpha\int\psi_e^*\psi_e d\mathbf{r} = 0 \quad (I.9)$$

since the total electron wavefunction is of course normalized.

The expectation value of the second neglected mixed term is proportional to

$$-\frac{\hbar^2}{2}\sum_\alpha\frac{1}{M_\alpha}\int\psi_e^*\nabla_\alpha^2\psi_e d\mathbf{r} = -\frac{\hbar^2}{2}\sum_\alpha\frac{1}{M_\alpha}\int\psi_e^*\nabla_i^2\psi_e d\mathbf{r}$$

$$= -\frac{\hbar^2}{2m_e}\sum_\alpha\frac{m_e}{M_\alpha}\int\psi_e^*\nabla_i^2\psi_e d\mathbf{r} \quad (I.10)$$

$$\simeq 0$$

where the last result obviously reflects the fact that $m_e/M_\alpha \ll 1$ for any possible nuclear mass value. We remark that in the formal development above, we have replaced the operator ∇_α^2 with the operator ∇_i^2 since it is reasonable to assume that $\psi_e^{(\mathbf{R})}(\mathbf{r}) = \psi_e(\mathbf{r} - \mathbf{R})$ and, therefore, to derive twice with respect to the $\mathbf{R}_\alpha$-variable is equivalent to derive twice with respect to the $\mathbf{r}_i$-variable. We also note that $\int \psi_e^* \nabla_i^2 \psi_e d\mathbf{r} \sim$ [electron kinetic energy].

In conclusion, the essence of the Born–Oppenheimer approximation consists in recognising that $m_e/M_\alpha \ll 1$ always and, accordingly, in setting to zero any term proportional to this ratio.

Further reading and references

[1] Colombo L 2021 *Solid State Physics: A Primer* (Bristol: IOP Publishing)
[2] Born M and Huang K 1954 *Dynamical Theory of Crystal Lattices* (Oxford: Clarendon)
[3] Böttger H 1983 *Principles of the Theory of Lattice Dynamics* (Weinheim: Physik-Verlag GmbH)
[4] Morrison M A, Estle T L and Lane N F 1976 Quantum States of Atoms *Molecules, and Solids* (Englewood Cliffs, NJ: Prentice-Hall)

IOP Publishing

Atomic and Molecular Physics (Second Edition)
A primer
Luciano Colombo

Appendix J

The one-dimensional quantum harmonic oscillator

Let us consider a one-dimensional harmonic oscillator with mass M. Its classical potential energy is $-\gamma x^2/2$, where x is the displacement from its equilibrium position and γ is the spring constant of the restoring force. The corresponding energy eigenvalue problem is

$$-\frac{\hbar^2}{2M}\frac{d^2u(x)}{dx^2} + \frac{1}{2}\gamma\, x^2\, u(x) = E\, u(x) \tag{J.1}$$

where $u(x)$ and E are, respectively, the quantum vibrational wavefunction and energy.

Equation (J.1) can be written in a more convenient form by using the dimensionless variables

$$\varepsilon = \frac{2E}{\hbar\omega_0} \quad \text{and} \quad \xi = \sqrt{\frac{M\omega_0}{\hbar}}\, x \tag{J.2}$$

where we have defined the fundamental oscillation frequency $\omega_0 = \sqrt{\gamma/M}$. By substitution into equation (J.1) we get the following differential equation

$$\left(\frac{d^2}{d\xi^2} - \xi^2 + \varepsilon\right)u(\xi) = 0 \tag{J.3}$$

for which we seek a solution of the form[1]

$$u(\xi) = g(\xi)\,\exp(-\xi^2/2) \tag{J.4}$$

By inserting this functional form into equation (J.3) we easily get

---

[1] This approach is motivated by considering the limiting situation $\varepsilon \ll \xi^2$. In this case the differential equation has the simplified form $(d^2/d\xi^2 - \xi^2)u(\xi) = 0$, which has solution $u(\xi) \sim \exp(\pm\xi^2/2)$.

doi:10.1088/978-0-7503-5734-0ch19

$$\frac{d^2 g(\xi)}{d\xi^2} - 2\xi \frac{dg(\xi)}{d\xi} + (\varepsilon - 1)g(\xi) = 0 \tag{J.5}$$

whose solution $g(\xi)$ is a finite polynomial in ξ, so that $u(\xi)$ remains finite; therefore, it must be $(\varepsilon - 1) = 2n$ with $n = 0, 1, 2, \ldots$.

The $g(\xi)$ function is usually taken in the form of an Hermite polynomial [1–4]

$$H_n(\xi) = (-1)^n \exp(\xi^2) \frac{d^n}{d\xi^n} \exp(-\xi^2) \text{ with } n = 0, 1, 2, 3, \ldots \tag{J.6}$$

and the corresponding *normalized wavefunctions for the one-dimensional harmonic oscillator* are written as

$$u_n(\xi) = N_n \exp(-\xi^2/2) \, H_n(\xi) \text{ with } N_n = (\sqrt{\pi} \, 2^n \, n!)^{-1/2} \tag{J.7}$$

which turn out to be either even (if n is even) or odd (if n is odd) under the inversion $x \to -x$ The energy eigenvalues are easily obtained

$$E_n = \left(n + \frac{1}{2} \right) \hbar \omega_0 \tag{J.8}$$

They correspond to a *non-degenerate discrete energy spectrum*. At variance with classical physics, the energy of a quantum oscillator can never be zero: its minimum value $\hbar \omega_0/2$ is referred to as the *zero-point energy*.

Further reading and references

[1] Sakurai J J and Napolitano J 2011 *Modern Quantum Mechanics* 2nd edn (Reading, MA: Addison-Wesley)
[2] Miller D A B 2008 *Quantum Mechanics for Scientists and Engineers* (New York: Cambridge University Press)
[3] Griffiths D J and Schroeter D F 2018 *Introduction to Quantum Mechanics* 3rd edn (Cambridge: Cambridge University Press)
[4] Bransden B H and Joachain C J 2000 *Quantum Mechanics* (Englewood Cliffs, NJ: Prentice-Hall)

Printed in the USA
CPSIA information can be obtained
at www.ICGtesting.com
JSHW061340241223
54197JS00004B/64